Gorana Rampazzo Todorovic

Behaviour of glyphosate and ampa in soils

Gorana Rampazzo Todorovic

Behaviour of glyphosate and ampa in soils

under the influence of different tillage systems and erosion

Südwestdeutscher Verlag für Hochschulschriften

Impressum / Imprint

Bibliografische Information der Deutschen Nationalbibliothek: Die Deutsche Nationalbibliothek verzeichnet diese Publikation in der Deutschen Nationalbibliografie; detaillierte bibliografische Daten sind im Internet über http://dnb.d-nb.de abrufbar.

Alle in diesem Buch genannten Marken und Produktnamen unterliegen warenzeichen-, marken- oder patentrechtlichem Schutz bzw. sind Warenzeichen oder eingetragene Warenzeichen der jeweiligen Inhaber. Die Wiedergabe von Marken, Produktnamen, Gebrauchsnamen, Handelsnamen, Warenbezeichnungen u.s.w. in diesem Werk berechtigt auch ohne besondere Kennzeichnung nicht zu der Annahme, dass solche Namen im Sinne der Warenzeichen- und Markenschutzgesetzgebung als frei zu betrachten wären und daher von jedermann benutzt werden dürften.

Bibliographic information published by the Deutsche Nationalbibliothek: The Deutsche Nationalbibliothek lists this publication in the Deutsche Nationalbibliografie; detailed bibliographic data are available in the Internet at http://dnb.d-nb.de.

Any brand names and product names mentioned in this book are subject to trademark, brand or patent protection and are trademarks or registered trademarks of their respective holders. The use of brand names, product names, common names, trade names, product descriptions etc. even without a particular marking in this works is in no way to be construed to mean that such names may be regarded as unrestricted in respect of trademark and brand protection legislation and could thus be used by anyone.

Coverbild / Cover image: www.ingimage.com

Verlag / Publisher:
Südwestdeutscher Verlag für Hochschulschriften
ist ein Imprint der / is a trademark of
AV Akademikerverlag GmbH & Co. KG
Heinrich-Böcking-Str. 6-8, 66121 Saarbrücken, Deutschland / Germany
Email: info@svh-verlag.de

Herstellung: siehe letzte Seite /
Printed at: see last page
ISBN: 978-3-8381-3469-7

Zugl. / Approved by: Vienna, BOKU, Diss., März 2012

Copyright © 2012 AV Akademikerverlag GmbH & Co. KG
Alle Rechte vorbehalten. / All rights reserved. Saarbrücken 2012

Behaviour of glyphosate and AMPA in soils under the influence of different tillage systems and erosion

Doctoral Thesis
of
Dipl.-Ing. MSc. Gorana RAMPAZZO TODOROVIC

Aimed academic degree:
Doktor der Bodenkultur
Doctor of Engineering Sciences
(Dr. rer. nat. techn.)

Immatriculation Number: 0340764
Study programme Number : H 088910
Field of Study: Soil Science
Supervisor: o.Univ.Prof.Dipl.-Ing. Dr.Dr.h.c.mult. Winfried.E.H. Blum

Vienna, March 2012

I dedicate this work to all whose love, understanding and support brought me so far.

First of all my parents Stanko and Mirjana, my brother Miroslav,

my husband Nicola and especially to the one who made it worth to be done,

my lovely son Darko Lorenzo

and last but not least all those who incoraged me on my way.

Contents

Abstract 5
Zusammenfassung 6

1 Introduction and aims of the study 9
2 Material and methods 13
   2.1 Investigated sites 13
   2.2 Herbicide application under agricultural field practice 15
   2.3 Rainfall simulation experiments 15
   2.4 Soil sampling and sampling preparation 20
   2.5 Laboratory methodology 21
      2.5.1 General soil physical parameters 21
      2.5.2 General soil chemical parameters 21
      2.5.3 General soil mineralogical parameters 22
      2.5.4 Development of a new method for the determination of glyphosate and AMPA with a HPLC-MS/MS 22

3 Results and discussion 28
   3.1 Investigated soils 28
      3.1.1 Soil profile descriptions 28
      3.1.2 Soil physical parameters 30
      3.1.3 Soil chemical parameters 31
      3.1.4 Soil mineralogical parameters 32
   3.2 A new method for the determination of glyphosate and AMPA with a HPLC-MS/MS 35
      3.2.1 Results of the verification of the extraction method 35
      3.2.2 Results of the validation study 39
      3.2.3 Results of the quality control 40
   3.3 Adsorption of glyphosate and AMPA in agricultural soils 43
   3.4 Rainfall simulation experimental results 48
   3.5 Dissipation of glyphosate and AMPA through natural erosion and leaching processes 61

4 Conclusions 65
5 References 70

Acknowledgments 76
Curriculum vitae... 77

Behaviour of glyphosate and AMPA in soils under the influence of different tillage systems and erosion

Gorana Rampazzo Todorovic
Institute of Soil Research
Department of Forest and Soil Sciences
University of Natural Resources and Life Sciences
Peter-Jordan-Straße 82, 1190 Vienna, AUSTRIA

Abstract

Although glyphosate is a heavily applied herbicide worldwide, the risk of environmental contamination through transport mechanisms of this substance is still not well documented.
Moreover, since sorption and desorption of glyphosate in soils is extremely pH dependent the use of the best suitable extraction solution is particularly relevant.
This work aimed to develop a new extraction method for an exact determination of glyphosate and AMPA in different soils, to investigate the time dependent behavior of glyphosate and AMPA in soils and to investigate the influence of soil erosion processes on the dissipation of glyphosate and AMPA.

The novel method is based on an extraction utilizing Na-tetraborate, an SPE clean-up step, and subsequent LC-MS/MS detection was developed. In the measurement procedure with LC-MS/MS, a negative mode was used and, FMOC derivatized glyphosate and AMPA ions were identified by the precise determination of their ion mass. The LOQ of both substances (13.8-22.7 µg kg^{-1} for glyphosate (RSD <10%) and 84.0-88.9 µg kg^1 (RSD <10%) for AMPA) in all investigated soil samples demonstrates the sensitivity of the method, which enables measurements of both substances in different soil matrixes. The success of the new extraction method was confirmed by measuring spiked soil samples to control recovery and detection limits. This was followed by applying the same procedure to different soil types with contrasting characteristics from the experimental fields after conventional glyphosate application. The developed method covers matrix effects of the most representative agricultural soil types of Austria but can be applied for glyphosate and AMPA investigations on any kind of mineral soils.

The method, which could precisely and reliably identify glyphosate and AMPA in soils even at low concentrations, is sensitive to co-extractable humic substances. Since the extraction is buffered with 40 mM tetraborate at pH 8.5, e.g. close to the zero point of charge (ZPC) of Fe-oxides and humic substances with variable positive electric charges, these soil constituents are co-extracted. Therefore the method focuses mainly to the glyphosate adsorption by Fe-oxides and organic matter.

The Pixendorf-chernozem features a low content, the Pyhra-stagnosol a medium content and the Kirchberg-cambisol a high content of Fe-oxides. Therefore, the expected sorption capacity for glyphosate and AMPA increased respectively from the chernozem, over the stagnosol to the higher weathered cambisol at Kirchberg.

Immediately after field application of Round Up Max glyphosate is first adsorbed in the upper 0-2 cm and is than transported and adsorbed in deeper horizons within few days with concomitant increase of the AMPA contents. The increase of AMPA 3 days after application of Round Up Max shows the very quick degradation of glyphosate to AMPA.

The results show that both the Kirchberg-cambisol and the Phyra-stagnosol, with a higher pedogenic iron-oxide content, adsorbed a distinctly higher quantity of glyphosate and AMPA than the Pixendorf–chernozem which had a distinctly lower iron-oxide content and a Kd-value about 10 times lower than the Kirchberg-cambisol. Thus, iron-oxides in general seem to be a key parameter for the glyphosate and AMPA adsorption in soils.

The rain simulation experiments could clearly show that even a potentially high erodable soil like the Pixendorf-chernozem suffers low damages, if adequate protection practices are applied, and respectively that even a potentially high adsorbing soil like the Kirchberg-cambisol suffers strong erosion damages, if its structure is unfavourable at the time of application. In this case in one of the experimental NoTillage-plot repetitions at Kirchberg up to 47 % of the applied glyphosate amount were dispersed with run-off.

Zusammenfassung

Obwohl Glyphosat weltweit stark eingesetzt wird, ist das Risiko einer Umweltkontaminierung durch Transportprozesse noch weitgehend ungeklärt. Da die Sorption und Desorption von Glyphosat in Böden extrem pH-abhängig ist, ist die Wahl der Extraktionslösung von entscheidender Bedeutung.

Ziele dieser Arbeit waren eine neue Extraktionsmethode für die exakte Bestimmung von Glyphosat und AMPA in verschiedenen Böden zu entwickeln, das zeitabhängige Verhalten in Böden sowie den Einfluss von Erosionprozessen auf die Verteilung von von Glyphosat und AMPA zu untersuchen.

Die neue Methode basiert auf eine Extraktion mittels Na-Tetraborat, eine SPE-Festphasenextraktion mit Aufreinigung und eine anschließende LC-MS/MS Messung. Bei der neuen Messtechnik mit LC-MS/MS wurde der negative Modus verwendet, wobei FMOC-derivatisierte Glyphosat- und AMPA-Ionen durch die präzise Bestimmung ihrer Ionenmasse detektiert werden. Die LOQ beider Substanzen (13.8-22.7 µg kg^{-1} für Glyphosat (RSD <10%) und 84.0-88.9 µg kg^{-1} (RSD <10%) für AMPA) zeigten in allen untersuchten Bodentypen die Empfindlichkeit der Methode, welche die Messung beider Substanzen in unterschiedlichen Bodenmatrizes ermöglicht. Der Erfolg der neuen Extraktionsmethode wurde durch Kontrolle der Wiederfindungsrate und der Detektionsgrenze der gemessenen dotierten Boden-Testproben bestätigt. Darauf aufbauend wurde die gleiche Vorgangsweise auf Bodenproben verschiedener Bodentypen mit unterschiedlicher chemisch-mineralogischer Zusammensetzung aus den Freilandversuchen nach der landwirtschaftlich üblichen Glyphosat-Applikation angewendet. Die Methode, welche Glyphosat und AMPA in Böden auch in geringen Konzentrationen präzise und zuverlässig identifizieren kann, ist empfindlich gegenüber koeluierender organischer Substanz. Da die Extraktion mit 40 mM Tetraborat-Puffer bei pH 8.5 gepuffert ist, d.h. nahe dem Ladungsnullpunkt (LNP) der meisten Fe-Oxide, und Huminstoffe mit variabler positiver elektrischer Ladung mitextrahiert werden, ist die Methode vor allem auf die Sorption von Glyphosat und AMPA durch Fe-Oxide und organische Substanz ausgerichtet.

Die neu entwickelte Methode berücksichtigt Matrixeffekte der repräsentativsten landwirtschaftlichen Böden Österreichs, kann aber an jeden Mineralboden zur Glyphosat- und AMPA-Untersuchung angewendet werden.

Der Pixendorf-Tschernosem weist einen niedrigen, der Pyhra-Pseudogley einen mittleren und die Kirchberg-Braunerde einen hohen Gehalt an Fe-Oxiden auf, daher stieg wie erwartet die Sorptionskapazität für Glyphosat und AMPA vom Tschernosem, über den Pseudogley zur stärker verwitterten Braunerde an. Unmittelbar nach der Round Up Max Applikation unter Freilandbedingungen wurde Glyphosat zuerst in den oberen 0-2 cm des Bodens sorbiert und wurde dann, bei gleichzeitiger Zunahme des AMPA Gehaltes, zu tieferen Bodenhorizonten innerhalb weniger Tage verlagert. Die Zunahme des AMPA Gehaltes 3 Tage nach Applikation von Round Up Max zeigt den sehr schnellen Abbau von Glyphosat zu AMPA.

Die Ergebnisse zeigten, dass sowohl die Kirchberg-Braunerde als auch der Phyra-Pseudogley, auf Grund der höheren Gehalte an pedogenen Fe-Oxiden, einen deutlich höheren Gehalt an Glyphosat und AMPA adsorbierten als der Pixendorf–Tschernosem mit weit niedrigerem Gehalt an Fe-Oxiden und einem ca. 10-fach geringeren Kd-Wert als die Kirchberg-Braunerde. Deshalb scheinen Fe-Oxide im Allgemeinen ein Schlüsselparameter für die Glyphosat- und AMPA Sorption in Böden zu sein.

Durch die Regensimulierung-Experimente konnte sehr deutlich gezeigt werden, dass auch potentiell stark erosionsgefährdete Standorte wie Tschernoseme aus Löß in Pixendorf geringe Schäden erfahren, wenn geeignete Erosionschutzmaßnahmen eingesetzt werden, bzw. dass auch in potentiell stark glyphosatsorbierenden Böden wie die Kirchberg-Braunerde enorme Erosionsschäden auftreten können, wenn der Bodenstrukturzustand und die hydraulischen Eigenschaften des Bodens ungünstig sind. In diesem Experiment wurden in Kirchberg in einer der beregneten No-Tillage-Parzelle bis zu 47 % der applizierten Glyphosat-Menge durch Oberflächenabfluß hangabwärts verlagert.

Keywords: Glyphosate / AMPA/ erosion / chernozem / cambisol / stagnosol / adsorption / HPLC / tandem mass spectrometry

Abbreviations:
AMPA - aminomethylphosphonic acid
FMOC - fluorenylmethyloxycarbonyl chloride
tetraborate buffer - Di-Natriumtetraborate-Decahydrate
RSD - relative standard deviation

SPE – solid phase extraction
LC-MS –liquid chromatography mass spectrometry
LC-MS/MS –liquid chromatography tandem mass spectrometry
LOD – limit of detection
LOQ – limit of quantification
Kd-value - adsorption coefficient value
RE – recoveries (%)
STD – standard variation

1 Introduction and aims of the study

There is growing concern about identifying and understanding the mechanisms that control the fate of chemicals as a source of environmental contamination, especially in soils and water. Organic pollutants are mainly man-made and industrially produced and nowadays they are present in water systems, run-off, soils and food (Ternan *et al.*, 1998), where soils play an important role in the regulation of contaminants in ecosystems.
A better understanding of the behavior of organic compounds in soils is important for the improvement of environmental protection. Therefore, there is a need for more specific pesticide management based on the adaptation of the pesticide type and application rates to the characteristics of the area of application (Peruzzo *et al.*, 2008).

The behaviour and attenuation of organic contaminants in soil environments are generally governed by a variety of climatic (e.g. rainfall distribution and –erosivity), geomorphological (e.g. relief, slope, run-off, and leaching), agricultural (uptake by plants) and pedological (physical, mineralogical, chemical and biological) processes, including precipitation, sorption, desorption, immobilization, volatilization, chemical and biological degradation, etc.) (Mamy *et al.*, 2005), see Fig. 1.

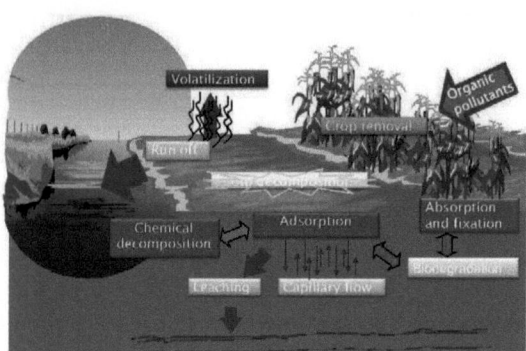

Figure 1: Dissipation ways of organic pollutants in soil (Mamy et al., 2005).

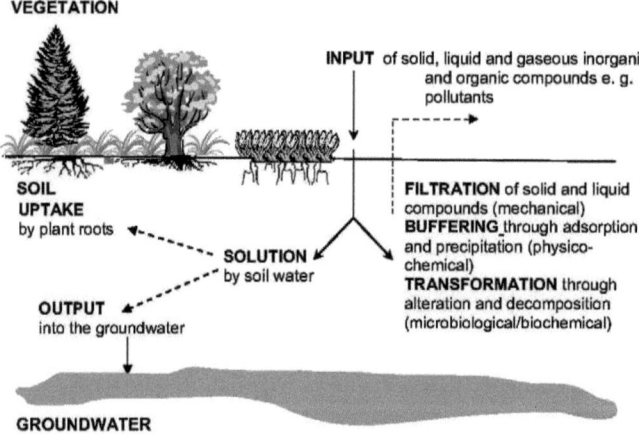

Figure 2: Behaviour of pollutants in the soil environment (Blum, 2007).

Organophosphonates are released into the environment in enormous quantities and, among the non-selective herbicides, glyphosate is applied at about 60% of the global sales (Candela et al., 2007). Glyphosate (N-(phosphonomethyl) glycine) is the active agent in Round Up Max, a post-emergency non-selective broad spectrum herbicide widely applied in agricultural practice. It blocks the shikimic acid pathway for biosynthesis of aromatic amino acids in plants, also in some microbes, but not in all soil microorganisms (Haghani et al., 2007). This herbicide is worldwide used in agriculture, railways, dam protection, urban areas (Ternan et al., 1998; Baylis, 2000). Glyphosate itself is an acid, but it is commonly used as a salt, most commonly as isopropylammonium salt.

Glyphosate is a polar, highly water-soluble substance that easily forms complexes with metals and binds tightly to soil components (Eberbach, 1997; Gimsing et al., 2004; Ghanem et al., 2007) but can be desorbed or leached out of the soil, thus, traces of this compound have been found in surface-water and groundwater systems (Landry et al., 2005; Peruzzo et al., 2008). The persistence of glyphosate is typically up to 170 days, with a half-life time of 45-60 days (Peruzzo et al., 2008). Some studies, however, also show a half-life time of years (Carlisle and Trevors, 1988; Zaranyika and Nyandoro, 1993, Eberbach, 1997). The major degradation product of glyphosate is aminomethylphosphonic acid (AMPA) (Peruzzo et al., 2008; Gimsing et al., 2004).

Like for many other organic contaminants, the glyphosate fate and behavior in soils is in principle affected by different specific soil factors and processes but depends also on the interaction between the herbicide and further specific local site conditions (Soulas and Lagacherie, 2001; Gimsing et al., 2004; Mamy et al., 2005; Eberbach, 1997). The investigation of the pathways of glyphosate and AMPA in the environment and their interaction with different soils is therefore mandatory for assessing their environmental impact. Consequently, farmers should strive for improved management practices in the use and release of these chemicals. This calls for considering the pesticide type and application rates as well as the characteristics of the application site (Soulas and Lagacherie, 2001).

The adsorption of glyphosate in soils is one of the major processes retaining glyphosate and AMPA against dissipation into ground- and surface water. It seems to be that particularly pedogenical Fe-(and Al-) oxides have a strong capacity to glyphosate adsorption through covalent bonds, (Mentler *et. al.*, 2007; Pessagno *et al.*, 2008; Barja and dos Santos Afonso, 1998; Barja *et al.*, 2001; Zhou *et al.*, 2004; Gimsing *et al.*, 2004; Morillo *et al.*, 2000; Piccolo *et al.*, 1994; Gerritse *et al.*, 1996; Eberbach, 1997).

Although glyphosate is a heavily applied herbicide worldwide, the risk of environmental contamination through transport mechanisms of this substance is still not deeply documented.

Therefore an estimation of time dependent degradation rates and consequently risks involved is very important, e.g. the glyphosate biodegradation at the beginning is generally fast, then slower with enhanced adsorption processes (Peruzzo *et al.*, 2008; Zaranyika and Nyandro, 1993).

Another risk for the dissipation of glyphosate and AMPA in the soil-water environment is the herbicide application at sites and soils susceptible to serious erosion processes. Since glyphosate is water soluble, there is a potential risk of run-off in case of erosive precipitations very soon after the herbicide application, before the plants could adsorb it and in the regions with high risk of run-off. Pesticide residues in the bottom sediments of the surface water systems may be influenced by a number of factors including the run-off potential and intrinsic properties of the pesticides. (Peruzzo *et al.*, 2008; Zaranyika and Nyandro, 1993).

Regarding the pathway of leaching-preferential flows, the sources and transport routes of pesticides to groundwater are multifaceted. In the study of Rueppel *et al.*, (1977), glyphosate was considered as pesticide I class and thereby posses no propensity for leaching. On the other hand, in certain soils and under different conditions, there is a risk for groundwater contamination as it is shown in some other investigations (Landry *et al.*, 2005).

One of the key problems for obtaining reliable results from field samples is controlling the possible biodegradation of glyphosate during storage and preparation of the soil samples, because the glyphosate content can be influenced by soil microbes (Zaranyika and Nyandro, 1993).

Another problem is the use of the best suitable extraction solution, since sorption and desorption of glyphosate in soils is extremely pH dependent (Morillo et al., 1997). In previous studies, different alkaline solutions were used to extract glyphosate and its metabolites from soils with different recovery rates (Borjesson and Torstensson, 2000; Ibáñez et al., 2005; Aubin and Smith, 1992 ; Thompson, 1989; Miles and Moye, 1988). Moreover, it is vital to adjust the concentration of the extraction media in such a way that high recovery rates can be obtained while avoiding matrix problems provoked by overly aggressive alkaline media, which enrich the dissolved humic substances in the extraction solution (Ibáñez et al., 2005; Sancho et al., 2002). The use of LC analysis is recommended for determining polar compounds and substances with low volatility, low mass and low concentration in soil, e.g. glyphosate (Ibáñez et al., 2005; Lee et al., 2002).

Considering all mentioned influences and interactions on the behavior of glyphosate and AMPA in soil environments, **the aims of this works were**:

1) To develop a new extraction method for an exact determination of glyphosate and AMPA in different soils even in very small quantities after extraction, SPE clean-up step based on FMOC-derivatization analytics and LC-MS/MS detection.

To validate the accuracy of the newel method by investigating the contents of glyphosate and AMPA after applying Round Up Max on complete different soil types.

2) To investigate the time dependent behavior of glyphosate and AMPA in soils. This was achieved by investigating the contents of glyphosate and AMPA after applying Round Up Max according to the agricultural practice on three different representative soil types in three different agricultural regions of Austria. The soils investigated were a cambisol (Kirchberg am Walde, Styria) covering approx. 50%, a chernozem (Pixendorf, Lower Austria) covering approx. 18% and a stagnosol (Pyhra, Lower Austria) covering approx. 10% of the agricultural land use of Austria, (Haslmayr, 2010).

Moreover, these three soil types were selected because of their contrasting physico-chemico-mineralogical parameters (e.g. texture, carbonate content, pH-value, and iron-oxide-content) for a better understanding of glyphosate behavior and extraction from soils. All three sites were under comparable tillage system (no-tillage and conventional tillage) in long-term experiments (Klik *et al.*, 2010).

3) To investigate the influence of soil erosion processes on the dissipation of glyphosate and AMPA after application of Round Up Max. This was achieved by two rain simulation experiments in the field conducted on two different locations in Austria (Kirchberg am Walder and Pixendorf) with different soil types. Both sites were under comparable tillage system (no-tillage and conventional tillage) in long-term experiments (Klik *et al.*, 2010). For this, Round Up Max was applied onto rain simulation soil plots according to the agricultural practice shortly before starting the rain simulation experiment (worst case scenario). After 60 minutes of rain simulation, run-off and soils samples were collected and analysed for glyphosate and AMPA.

2 Material and Methods

2.1 Investigated sites

Kirchberg am Walde (Styria)

Site description:

The experiment was carried out at agricultural experimental fields, where different tillage systems (no-tillage NT= direct drill, no plough, with a winter green vegetation cover and mayze crop in spring, and conventional tillage = CT, plough with or without a winter green vegetation cover) in 3 field replications are tested since 2007.
Geographic coordinates: 48° 16' N and 15° 58' E
Yearly average precipitation: 730 mm
Yearly average temperature: 9,1° C
Slope : 12-15 %
Soil type : stagnic Cambisol (WRB 2006)

Pyhra (Lower Austria)
Site description:

The experiment was carried out at agricultural experimental fields, where different tillage systems (no-tillage NT= direct drill, no plough, with a winter green vegetation cover and mayze crop in spring, and conventional tillage = CT) in 3 field replications are tested since 1999.
Geographic coordinates: 48° 08' N and 15° 42' E
Yearly average precipitation: 945 mm
Yearly average temperature: 9,4° C
Slope : 14-16 %
Soil type : Stagnosol (WRB 2006)

Pixendorf (Lower Austria)
Site description:

The experiment was carried out at agricultural experimental fields, where different tillage systems (no-tillage NT= direct drill, no plough, with a winter green vegetation cover and mayze crop in spring, and conventional tillage = CT) in 3 field replications are tested since 1999.
Geographic coordinates: 48° 16' N and 15° 58' E
Yearly average precipitation: 685 mm
Yearly average temperature: 9,4° C
Slope : 5-6 %
Soil type : Chernozem (WRB 2006)

2.2 Herbicide application under agricultural field practice

The Round Up Max application was performed at all three sites according to the common agricultural practice, i.e. 4 liters Round Up Max (450 g glyphosate /L Round Up Max) were dissolved in 200 liters of water and applied per ha (2 % herbicide solution). This corresponds to an application of 1800 g glyphosate ha^{-1} or 180 mg glyphosate m^{-2}. The application was carried out at sunny and not windy weather at the no-tillage (NT) plots.

2.3 Rainfall simulation experiments

For a better understanding of the influence of erosion processes on the behaviour of glyphosate and AMPA and their possible dispersion under rainy conditions shortly after application in the practice, two rain simulation experiments were conducted on two different locations in Austria (Kirchberg am Walde, Styria and Pixendorf, Lower Austria) with different soil types (cambisol at Kirchberg and chernozem at Pixendorf). These locations are the same experimental fields as above, where glyphosate is usually applied according to agricultural practices.

Both soils have a rather high erosion risk due to their texture (fine sandy in Kirchberg and silty in Pixendorf), as it could be observed during field excursions, see Fig. 3, so erosion processes yearly lead to loss of organic matter, nutrients and clay downslope, (Strauss, 2008).

Figure 3: Erosion rills at Kirchberg (left) and Pixendorf (right) under corn crop.

Another main difference between the two soils consists in the amount of pedogenic iron-oxides (Kirchberg features 16.000 mg dithionite-extractable (Fed) Fe . kg^{-1} soil and Pixendorf about 8.000 mg (Fed) Fe . kg^{-1} soil), and pH (acidic at Kirchberg and slightly alkaline at Pixendorf).

The rain simulation experiments took place in 3 field replications (1, 2, 3) within the Conventional Tillage (CT)- and the No Tillage (NT)- plots. The average slope in both sites was 12-15 % at Kirchberg and 5-6 10% at Pixendorf. Before starting the rain simulation, erosion plots were installed inside of each field replication in a dimension of 2m x 2 m, see Fig. 4.

Figure 4: Installation of the 2m x 2m rain simulation plot.

The culture type at time of the experiments was different in both sites (Kirchberg stood immediately after the mais yield in October whereas Pixendorf had a green cover after the wheat yield of July) but in both sites the vegetation cover degree was distinctly higher in the NT-plots than in the CT-plots, see Fig. 5-6.

Figure 5: Rain simulation plots at Kirchberg, CT-plot (left), NT-plot (right).

Figure 6: Rain simulation plots at Pixendorf, CT-plot (left), NT-plot (right).

Before starting the rain simulation experiments the water containers were filled up with deionizied water. The deionization was carried out by washing through 2 exchange clearing up tanks and controlling the electrical conductivity, see Fig. 7.

Figure 7: Deionization of the rain simulation water throuh clearing up tanks (left) and control of its electrical conductivity (right).

Following the common agricultural practice, for the rain simulation experiments a 2 % Round Up Max solution was prepared and sprayed homogeneously by a hand pump in the same concentration and amounts as in practice, e.g. 180 mg glyphosate / m^2 respectively 720 mg glyphosate / simulation plot, see Fig. 8. Immediately after application of Round Up Max the rain simulation started (worst case scenario) and lasted 60 minutes with an amount of 30 mm, see Fig. 8.

Figure 8: Application of the Round Up Max solution (2%) (left) and rain simulation (right).

The rain simulator was designed as a portable out- and indoor equipment. The spray pattern got generated by full jet nozzles. The rainfall intensity was controlled with intermittent spraying. The opening cycles of the solenoid valves were fully programmable with computing equipment (Strauss *et al.*, 2000).

The crucial elements for the construction of the simulator were:

Nozzles: ½ HH30WSQ, ½ HH40WSQ from Spraying System, USA.

Solenoid valves: 220-V, 3/2 way, model 131 from Bürkert, Germany.

Pressure regulators: P16A from Wilkerson Corp., USA.

Water suction pumps: Art. Nr. 0830 from Semadeni, Ostermundingen, Switzerland.

Steering Interface and Steering Programme, self-made equipment.

Drop size distribution: median volumetric size 2,1mm

Kinetic energy: 17 J/mm

2.4 Soil sampling and preparation

<u>Soil sampling for general soil parameters :</u>

Soil bulk samples from all plots (NT and CT) were taken for physico-chemico-mineralogical analysis at each site at two soil depths (0-5 cm, 5-20 cm), collected from 10 different points / field replication . The samples were air-dried and sieved at 2 mm size (fine earth). Moreover, for further physical analysis undisturbed samples (cylinders with 200 cm volume) were taken separated from each NT and CT field replication at 5-15 cm soil depth each in 5 repetitions.

<u>Soil sampling for the glyphosate and AMPA determination after herbicide application in the no-tillage (NT)-experimental fields according to agricultural practice:</u>

In order to investigate the fate of glyphosate and AMPA in depth and time after Round Up Max application, soil bulk samples were taken at different time intervals after application at 10 points within each no-tillage (NT)-field replication (pooled than to one sample per site) as follows:

Kirchberg :
a) immediately after the Round Up Max application, at 0-2 cm soil depth;
b) 3 days after application at 0-2 cm and 2-5 cm soil depth;
c) 12 days after application at 0-2 cm, 2-5 cm and 5-10 cm soil depth;

Pyhra:
a) immediately after the Round Up Max application, at 0-2 cm soil depth;
b) 28 days after application at 0-2 cm, 2-5 cm and 5-10 cm soil depth;

Pixendorf:
a) immediately after the Round Up Max application, at 0-2 cm soil depth;
b) 3 days after application at 0-2 cm and 2-5 cm soil depth;
c) 10 days after application at 0-2 cm, 2-5 cm and 5-10 cm soil depth;

After each soil sampling soil samples were immediately transported to the laboratory in cooling boxes. In the laboratory all samples were stored at -18 °C until measurements.

<u>Soil and run-off sampling for the glyphosate and AMPA determination after herbicide application and rain simulation experiments in the no-tillage (NT)- and conventional tillage (CT)-rain simulation plots:</u>

During the rain simulation, run-off-fractions were collected at different time intervals and cooled in boxes. Immediately after the rain simulation, soil samples were collected within the simulation soil plots at different depths (0-2 cm, 2-5 cm, 5-10 cm, 10-15 cm, 15-20 cm at Pixendorf and at 0-2 cm, 2-5 cm at Kirchberg). The run-off-samples as well as the soil samples were immediately transported to the laboratory in cooling boxes. There the run-off samples were centrifuged in order to separate the liquid from the solid phase.

2.5 Laboratory methodology

2.5.1 General soil physical parameters

- Particle size distribution, (wet sieving and pipette-methode), after ÖNORM L 1061;
- Dry bulk density and total porosity after ÖNORM L 1068-88;
- Water contents (Vol%), pore size distribution and available field capacity after KRENN (1999);

2.5.2 General soil chemical parameters

- pH-value ($CaCl_2$) after ÖNORM L 1083-89;
- Total carbon content after ÖNORM L 1080-1089.
- $CaCO_3$-content, (SCHEIBLER-method) after ÖNORM L 1084-89;
- Humus content, calculated from the Corg-content after ÖNORM L 1080-89;
- Oxalate soluble Fe (Feo) after Schwertmann (1964);
- Pyrophosphate soluble Fe (Fep) after Bascomb (1968);
- Dithionite soluble Fe (Fed) after Holmgren (1967);

2.5.3 General soil mineralogical parameters

- Semiquantitative mineral distribution by x-ray-diffractometry (Cukα-radiation) after SCHULTZ, (1964);
- Semiquantitative mineral distribution by x-ray-diffractometry (Cukα-radiation) after BRINDLEY &.BROWN, 1980; RIEDMÜLLER, 1978.

The frozen fine earth samples were directly dried using a freeze dryer from Christ Alpha (-55 ° C, 0.22 mbar) for the subsequent necessary homogenization and sieving steps.

2.5.4 Development of a new method for the determination of glyphosate and AMPA with a HPLC-MS/MS

Chemicals

Glyphosate, pestanal grade, and AMPA (99%) reference standards were obtained from Sigma-Aldrich Co (Vienna, Austria). Isotope labeled glyphosate 1,2-^{13}C, ^{15}N (100 ngμL^{-1}, 1.1 ml) and AMPA ^{13}C, ^{15}N (100 ngμL^{-1}, 1.1 ml) from Dr. Ehrenstorfer GmbH (Augsburg, Germany) were used as internal standards. FMOC Chloride (FMOC), puriss. p.a., derivatization grade (for HPLC,) % (HPLC) was purchased from Sigma-Aldrich (Vienna, Austria). Analytical grade Ethylendiamintetraacetic acid tetrasodium salt hydrate, 98%, was obtained from Alfa Aesar GmbH & Co KG (Karlsruhe, Germany). Di-Natriumtetraborate-Decahydrate (tetraborate buffer), potassium hydroxide (KOH), p.a., hydrochloric acid (HCL), 37 %, ACS, ISO, Reag. Ph Eur were from Merck KGaA (Darmstadt, Germany) as well as LC grade methanol (MeOH) and dichlormethane. Water and acetonitrile of LC/MS grade were from Fisher Chemie (Leicestershire, UK). Formic acid, rotipuran,≥98 %, p.a., ACS was obtained from Carl Roth GmbH + Co KG (Karlsruhe, Germany). For the mobile phase preparation, ammonium acetate, puriss. p.a. ACS from Fluka, Chemical Company Ltd., (Buchs, Switzerland) (NH$_4$Ac) was used.

Ultrapure water was obtained by purifying in two steps, first with Millipore Elix 5 and then with Millipore Simplicity 185 (Billerica, Ma, USA).

Separated standard stock solutions were prepared by dissolving accurately weighed glyphosate and AMPA reference standards in water each in concentrations of 50 µg L^{-1} to 2000 µg L^{-1}. Isotopically labeled standards were prepared by dissolving 60 µg L^{-1} of 1,2-^{13}C, ^{15}N glyphosate and ^{13}C, ^{15}N AMPA in water. For the extraction and derivatization steps, a solution of 40 mM tetraborate buffer dissolved in water, pH 8.5, was used. 6.5 mM FMOC-Cl was dissolved in LC/MS grade acetonitrile (ACN). 1 M EDTA solution was used for complexation of metal ions in the soil extracts. In the derivatization procedure, 6 M KOH and HCl, as well as 0.1 % formic acid were used. Ammonium hydroxide ammoniac solution 33% "Baker" was obtained from J.T. Baker (Deventer, The Netherlands).

Instrumentation

For analysis of glyphosate and AMPA in the soil extracts, an Agilent 1200 HPLC – system was combined with an Agilent 6410 Triple Quad system from Agilent Technologies Inc. (Santa Clara, CA, USA). The principle of MS/MS technique is a maximum of selectivity and a minimum of matrix detection. A tandem mass spectrometer reduces chemical noise for low-level quantification in a "dirty" soil matrix.

HPLC and MS parameters

Tab. 1 and 2 show the HPLC and ESI conditions.

Table 1: HPLC conditions.

Parameter	Setting
Analytic column	Agilent Zorbax Eclipse C-8 XDB (2.1 x 150 mm, 5 μm)
Column guard	Agilent Zorbax Eclipse C-8 XDB (2.1 x 12.5 mm, 5 μm)
Solvent A	99% (5 mM NH_4Ac) pH 8.5 / 1% ACN
Solvent B	1% (5 mM NH_4Ac) pH 8.5 / 99% ACN
Flow	0.3 mL min^{-1}
Column temperature	40 °C
Injection volume	3 μL
Run time	20 min

Table 2: ESI conditions.

Drying gas temperature	325 °C
Drying gas flow	8L min^{-1}
Nebulizer pressure	40 psi
Capillary voltage (negative mode)	4000 V

The mobile phase consisted of 5 mM ammonium acetate (NH_4Ac) dissolved in LC-MS grade water of pH 8.5, adjusted with (25%) ammoniac, and then prepared with LC-MS grade ACN. The pH-value of the buffer was controlled with a pH-electrode.

The separation took place through an Agilent 1200 Series binary HPLC-system (Santa Clara, CA, USA). Tab. 3 shows the gradient time-table of the solvents used.

Table 3: Gradient time table of the solvents A 99% ((5 mM NH$_4$Ac) pH 8.5 / 1% ACN) and B 1% ((5 mM NH$_4$Ac) pH 8.5 / 99% ACN).

Time	A%	B%
0	97.5	2.5
0.5	97.5	2.5
5.5	5.0	95.0
14.5	5.0	95.0
15.0	97.5	2.5
18.0	97.5	2.5

In order to avoid a contamination of the mass spectrometer by ionic compounds and remnants of the derivatization substances, the eluent was by-passed into a waste collection receptacle through a control valve during the time of 1.7-2.6 and 8.7-14 min.

In Tab. 4 the retention times of the compounds investigated are presented.

Table 4: Retention times of the investigated compounds.

Compound name	Retention time (RT)
FMOC-GLY (C13 N15)	4.978
FMOC-GLY	4.944
FMOC-AMPA (13 N15)	5.810
FMOC-AMPA	5.824

Tab. 5 shows the MS/MS, MRT conditions. MS/MS conditions were 90 V fragmentation voltage and 5 V collision energy.

Table 5: MS/MS, MRT conditions.

Compound name	Precursor ion	MS1 resolution	Product ion	MS1 resolution	Dwell time (ms)	Frag (V)	CE (V)
FMOC-GLY (C13 N15)	393	Wide	171	Unit	40	90	5
FMOC-GLY	390	Wide	168	Unit	40	90	5
FMOC-AMPA (13 N15)	334	Wide	112	Unit	40	90	5
FMOC-AMPA	332	Wide	110	Unit	40	90	5

Development of the extraction method

In developing the extraction method, an aliquot of 3 g of the certified soil reference material Eurosoil 026, Gawlik and Muntau (1999) were transferred into 50 ml centrifuge tubes. The soil parameters of the soil reference material used (Eurosoil 026) are shown in Tab. 6.

Table 6: Soil parameters of the certified soil reference material Eurosoil 026.

CaCO$_3$ (%)	Corg (%)	pH (CaCl$_2$)	C/N	Sand (%)	Silt (%)	Clay (%)	Fed * mg.kg^{-1} soil	Feo ** mg.kg^{-1} soil	CEC (mmol IE. kg^{-1} soil)
<0.5	1.05	4.6	21.0	64.3	31.1	4.6	12600	3800	9.8

* Fed = Dithionite-extractable Fe
**Feo = Oxalate-extractable Fe

The soil samples were spiked with 1 ml of 500 µg L^{-1} standard solutions of glyphosate and AMPA and shaken for 2 h on the mechanical shaker. Thereafter, to test different extraction solutions, 0.1 M KOH and 5% mM tetraborate buffer extraction solutions were added to the spiked soil samples in five replicates for both extraction solutions.

Owing to the higher chromatographic efficiency obtained for samples extracted with tetraborate buffer, we decided to continue the development of the extraction method with this compound. KOH extracts more humic acids and humic substances and increase matrixes effects. Humic substances influence the efficiency of the chromatographic column, probably based on the interaction with the stationary phase of the column, and stress the transfer capillary of the mass-spectrometer with high molecular matrix masses. Since 5% tetraborate buffer is a rather high extracting concentration, we minimized the ion concentration by reducing the tetraborate buffer concentration to 40 mM, which is the concentration of the derivatization buffer for FMOC derivatization. Again, based on the results obtained, we concluded that 40 mM tetraborate buffer is the more suitable extraction solution. To prove the suitability of the chosen extraction solution, the recovery rates from the spiked soil after incubation with three different glyphosate concentrations (100, 200 and 500 µg L^{-1}) were measured. To test this method on different soils, soils from the experimental sites at Pixendorf, Phyra and Kirchberg were sampled.

In order to avoid possible microbial degradation of glyphosate and AMPA in the soil samples, they were put in refrigerator boxes immediately after sampling and transported to the laboratory. In the laboratory all samples were stored at -18 °C. Since the standard procedure for preparing soil samples is air drying, homogenizing and sieving with a 2 mm sieve, it would normally be necessary to first thaw the samples. In order to avoid possible microbial degradation during the thawing and air drying period, the frozen samples were directly dried using a freeze dryer (-55 °C, 0.22 mbar) for the subsequent necessary homogenization and sieving steps. A Christ Alpha (Osterode, Germany) freeze dryer was used for freeze drying. The extracts of the soil samples were centrifuged using a Multifuge S Haereus (Hanau, Germany).

Based on the method described above, the freeze-dried soil samples were homogenized and an aliquot of 3 g sub-samples were transferred to 50 mL centrifuge tubes. To extract glyphosate and AMPA we used 30 ml 40 mM tetraborate buffer (pH 8.5). The samples were shaken for 1 h on the mechanical end-over-end shaker, and then centrifuged at 3452 g with 4200 rotations.min^{-1} for 25 min.

Derivatization method

After that, the liquid extract was sampled with a plastic syringe and filtered (nylon membrane filter 0.45 µm). An aliquot of 20 mL was transferred to a 250 mL plastic flask. The sample amount was registered gravimetrically. The samples were then acidified with 6 M HCl to pH 1.2, briefly stirred in a Vortex-shaker (Bohemia, N.Y., USA) and incubated for 30 min at room temperature in order to equilibrate the sample. Afterwards, we added 100 µL isotope spike-solution of ^{13}C and ^{15}N labelled glyphosate and AMPA (spike-concentration: 60 µg L^{-1}). The closed flask was shaken well and, for equilibration, allowed to stand for 60 min. Then the sample was neutralized by adding 6 M KOH. The sample was again intensively shaken and allowed to stand for a further 30 min, before 10 mL 40 mM borate buffer and 60 mL highly pure water from Millipore (Billerica, Ma, USA) were added. Subsequently, 10 mL of the organic FMOC-Cl solution (6.5 mM) were added, the sample was again shaken and incubated for 15 h at room temperature (20°C). The reaction was then stopped by adding 1 mL concentrated formic acid at about pH 3.0 (control by pH-electrode). Then, 4 mL 0.1 M EDTA were added to the solution mixture in order to avoid a complexation of the derivates. The sample was diluted with 100 mL highly pure water in order to reduce the amounts of organic dilutants down to 5% in relation to the subsequent SPE-enrichment.

SPE procedure

After the derivatization, the soil samples were cleaned up with strata-X 33u Polymeric Reversed Phase from Phenomenex (Aschaffenberg, Germany) 200 mgg^{-1} (strata-X) cartridges. The strata-X cartridges were placed at the vacuum manifold of Supelco, Sigma-Aldrich Comp. (St. Louis, MO, USA) and conditioned according to the strata-X manual.

A 70 mL reservoir with inserted frit was connected with the cartridge by an adapter. The reservoir was filled with the sample (flow speed of max. 2 mL min^{-1} is adjusted with the vacuum manifold). The eluate was collected in PE-flasks (50 mL) and discarded. After drying, SPE-cartridges were carefully flushed with 3.5 mL dichlormethane (HPLC-grade) and the eluate was collected in small glasses and discarded. The columns were again drawn dry at maximum vacuum; subsequently, the analytics were eluted from the columns with 9 mL MeOH (HPLC-grade) and collected in glass vials (20 mL).

The 9 mL SPE eluate was pre-concentrated to a volume of 200-300 µL at a temperature of 50°C using a Techne Sample Dri-Block DB-3A (Cambridge, UK) heating unit. The organic solvent was volatilized under weak nitrogen flow with a Techne Sample Concentrator (Cambridge, UK)-blow off unit. The concentrated samples were transferred to a 1.5 mL glass vial and the original recipient was flushed three times with 100 µL MeOH (HPLC-grade); the flush solutions were added to the concentrated sample. The samples were reduced to near dryness and subsequently dissolved in 250 µL of 90% 5 mM aqueous NH$_4$Ac (pH 8.5) and 10% ACN (LC-MS Grade).

3 Results and discussion

3.1 Investigated soils

3.1.1 Soil profile descriptions

Kirchberg am Walde (Styria)

The soil was classified as stagnic Cambisol (WRB, 2006) from tertiary carbonate free sediments with the following profile description, after Nestroy *et al.*, 2000:

Ap (0 – 25 cm): loamy sand, skeleton 3%, Munsell (moist) 10 YR/4/2, medium humous, no carbonate, crumby structure, medium porous, slightly acidic pH, rooted, few earthworm channels, abrupt boundary to:

BP (25 – 50 cm): loamy sand, skeleton 3%, Munsell (moist) (10 YR 5/4), weakly humous, no carbonates, polyhedral structure, medium porous, acidic pH, few rusty brown and pale dots, single Fe-concretions, well rooted, few earthworm channels, gradual boundary to:

Cv (50–200 cm): loamy sand, Munsell (moist) (10 YR 5/8 bis 2,5 Y), no carbonates,

coherent structure, weakly porous, some Fe-concretions, not rooted, no earthworm channels.

Pyhra (Lower Austria)

The soil was classified as Stagnosol (WRB, 2006) from carbonate free sediments (flysch, sandstone) with the following profile description, after Nestroy *et al.*, 2000:

Ap (0 – 25 cm): sandy loam, skeleton < 1%, Munsell (moist) 10 YR/3/2, strongly

	humous, no carbonate, coarse crumby structure, medium porous, slightly acidic pH, strongly rooted, few earthworm channels, gradual boundary to:
P (25 – 50 cm):	loam, skeleton < 1%, Munsell (moist) (10 YR 6/2), medium humous, no carbonates, cracky blocky (dry phase)/coherent (wet phase) structure, weakly porous, acidic pH, single Mn-concretions, poorly rooted, few earthworm channels, gradual boundary to:
S (50–200 cm):	loamy clay, skeleton < 1%, Munsell (moist) (10 YR 6/4), no carbonates, cracky blocky (dry phase)/coherent (wet phase) structure, no coarse and medium pores, many pale and rusty spots, rusty Fe-concretions, not rooted, no earthworm channels.

Pixendorf (Lower Austria)

The soil was classified as Chernozem (WRB, 2006) from loess with the following profile description after, Nestroy *et al.*, 2000:

Ap (0 – 25 cm):	sandy Silt, skeleton < 1%, Munsell (moist) 10 YR/3/2, medium humous, high carbonate content, medium crumby structure, medium porous, slightly alkaline pH, strongly rooted, many earthworm channels, gradual boundary to:
ACv (25 – 50 cm):	silty Loam, skeleton < 1%, Munsell (moist) (10YR/6/2), weakly humous, high carbonate content, subpolyherical structure, medium porous, slightly alkaline pH, rooted, pseudomycellium-like carbonate concretions, few earthworm channels, gradual boundary to:
Cv (50–100 cm):	Silt, skeleton < 1%, Munsell (moist) (10 YR 7/4), no humus, high carbonate content, coherent structure, few fine pores, no earthworm channels.

3.1.2 Soil physical parameters

Tab. 8 and 9 show the main soil physical parameter of the 3 investigated soils.

Table 8: Particle size distribution and texture of the investigated soils.

Site	soil type (WRB)	depth cm	texture	sand %	silt %	clay %
Pixendorf	chernozem	0-5	sandy silt	23.6	64,9	11.5
		5-20	sandy silt	22.7	64.5	12.7
Kirchberg	cambisol	0-5	loamy sand	52.8	33.2	14.0
		5-20	loamy sand	53.6	32.2	14.2
Phyra	stagnosol	0-5	Loam	34.9	39.9	25.2
		5-20	Loam	32.5	47.3	20.2

The investigated soils show their distinguished formation through very different textures, see Tab. 8. The Pixendorf-chernozem shows the development from loess with a typical high silt content. The Pyhra-stagnosol has a heterogeneous texture, typical for a loamy soil and the high amount of silt and clay explains the water stagnation of this soil. The Kirchberg-cambisol is a sandy soil, with > 50 weight % sand fraction.

Table 9: Bulk density (dB), total porosity (TP), pore size distribution and available field capacity (aFC) the investigated soils at the CT and NT plots.

Site	soil type (WRB)	plot	depth cm	dB $Mg.m^{-3}$	TP vol%	cP* vol%	ncP* vol%	mP* vol%	fP* vol%	aFC* mmWC
Pixendorf	chernozem	CT	5-15	1.51	43.2	12	5	15	11	20
		NT	5-15	1.62	38.8	9	4	15	12	19
Kirchberg	cambisol	CT	5-15	1.62	38.8	8	6	15	9	21
		NT	5-15	1.74	34.3	5	4	16	9	20
Phyra	stagnosol	CT	5-15	1.47	44.5	10	6	16	12	21
		NT	5-15	1.65	39.2	7	4	15	13	19

cP* = coarse pores (> 50µm)
ncP* = narrow coarse pores (50-10 µm)
mP* = medium pores (10-0.2 µm)

fP* = fine pores (< 0.2 µm)
aFC* = available field capacity in mm water column (WC)
There is a distinguished difference between the CT-plots and the NT-plots. NT-plots show higher bulk densities and lower total porosity than conventionally tillaged (CT) plots. This compaction is known from the literature, as e.g. Rampazzo et al., (1999) showed, and is due to a natural settlement of particles free from tillage practices. As a consequence, there is a loss of coarse pores which leads to a slightly diminiushed available water capacity, see Tab. 9.

3.1.3 Soil chemical parameters

Tab. 10 shows the main soil chemical parameter of the 3 investigated soils.

Table 10: General chemical parameters of the investigated soils.

Site	soil type (WRB)	depth cm	Nt %	Ct %	CaCO$_3$ %	Canorg %	Corg %	OM %	pH CaCl$_2$	C/N
Pixendorf	chernozem	0-5	0.13	3.7	14.7	1.80	1.86	3.2	7.3	14.8
		5-20	0.11	3.1	15.4	1.85	1.25	2.2	7.3	11.4
Kirchberg	cambisol	0-5	0.13	1.7	<0.5	0.0	1.73	3.0	5.7	13.3
		5-20	0.11	1.5	<0.5	0.0	1.51	2.6	5.8	13.7
Phyra	stagnosol	0-5	0.12	1.6	<0.5	0.0	1.61	2.8	5.7	13.4
		5-20	0.10	1.2	<0.5	0.0	1.17	2.0	5.6	12.2

The silty Loess-chernozem at Pixendorf is slightly alkaline with a medium carbonate content. The siliceous sites Pyhra and Kirchberg are weakly acidic, according to their formation conditions. The contents of soil organic matter decrease with soil depth, see Tab. 10.

3.1.4 Soil mineralogical parameters

Pedogenic Fe-(Al)-oxides are important indicators for e.g. weathering intensity and redox processes in soils. They play also a major role as adsorbers for glyphosate in soils. The dithionite soluble Fe indicates the total amount of pedogenic formed Fe-oxides, that means organically bound, amorphous and well crystallized. This Fe-fraction does not allow any conclusion about the type of Fe-oxide (goethite is however the most relevant Fe-oxide in Central Europe).

The Pixendorf-chernozem features a rather low content, the Pyhra-stagnosol a medium content and the Kirchberg-cambisol a high content of Fe-oxide, see Tab. 11. From this the expected sorption capacity for glyphosate and AMPA increases respectively from the chernozem, over the stagnosol to the higher weathered cambisol at Kirchberg.

Table 11: Fe-oxide distribution in the investigated soils.

Site	soil type (WRB)	depth cm	Fep* mgkg^{-1}	Feo** mgkg^{-1}	Fed*** mgkg^{-1}	Feo/Fed
Pixendorf	chernozem	0-5	37	983	7970	0.12
		5-20	39	1040	8378	0.12
Kirchberg	cambisol	0-5	530	3422	14843	0.23
		5-20	569	3726	15032	0.25
Phyra	stagnosol	0-5	550	3241	9959	0.32
		5-20	538	3215	9918	0.32

Fep* = organic bound Fe-oxides, pyrophosphate-soluble
Feo** = „amorphous" (weakly crystallized) Fe-oxides, oxalate-soluble
Fed*** = well crystallized Fe-oxides , dithionite-soluble

All sites show a low Feo/Fed ratio which indicates that in all soils investigated the most relevant amount of pedogenic Fe-oxides is well cristallized, see Tab. 11.

Tab. 12 shows the mineral distribution of the soils under investigation.

Table 12: Semiquantitative total mineral distribution (mass %) in the investigated soils.

Site	soil type (WRB)	depth cm	micas %	quartz %	feldspars %	calcite %	dolomite %	pyrite %
Pixendorf	chernozem	0-20	47	24	12	10	7	0
Kirchberg	cambisol	0-20	42	31	27	0	0	0
Phyra	stagnosol	0-20	45	37	18	0	0	0

All investigated soils are characterized by the dominance of layer silicates mostly micas. Quartz is detectable in all soils, so feldspars, but less in the carbonatic Pixendorf-chernozem. Carbonates (calcite and dolomite) are represented in the Loess chernozem as expected, whereas the siliceous Phyra-stagnosol and Kirchberg-cambisol are free of carbonates in concomitance with their genetical formation, see Tab. 12.

All investigated soils show a distinguished high content of expandable clay minerals (smectite and vermiculite), especially in the Pyhra-stagnosol where they amount up to > 50%, which explains the stagnic characteristics of that soil. Illite was found in all soils but with higher amounts in the cambisol and stagnosol. Kaolinite is a typical indicator for acidic conditions and shows the highest contents in the acidic Kirchberg-cambisol. Chlorite was detected in all soils, but it must be distinguished between a „primary" chlorite as a mineral of Loess in the chernozem, and a „secondary" chlorite (pedogenetically formed) in the stagnosol and in the cambisol. In all soils traces of mixed layers minerals were found, see Tab. 13.

Table 13: Semiquantitative clay mineral distribution (mass %) in the investigated soil.

Site	soil type (WRB)	depth cm	smectite %	vermiculite %	illite %	kaolinite %	chlorite %	mixed layers %
Pixendorf	chernozem	0-20	20	11	49	10	10	Tr.
Kirchberg	cambisol	0-20	17	8	48	24	3	Tr.
Phyra	stagnosol	0-20	27	28	34	8	3	Tr.

3.2 A new method for the determination of glyphosate and AMPA with HPLC-MS/MS

3.2.1 Results of the verification of the extraction method

The three investigated soil differed especially in soil texture, amount of pedogenic iron-oxides (Fed), pH-value, and the amount of extractable humic substances from the carbon pool, which have the strongest influence on the extraction procedure for glyphosate and AMPA. Therefore, the matrix effect of the Kirchberg soil was more pronounced because of a higher content of extractable humic substances.

The first step of the method development aimed at optimizing the soil homogenization procedure, avoiding biodegradation of the investigated analytes. A freeze dryer helps avoid potential glyphosate and AMPA losses during the drying period at room temperature in the laboratory.

The next step involved selecting the extraction solution for extraction of glyphosate and AMPA from soil samples. Based on previous studies, e.g. Ibáñez et al., (2005), two different alkaline extract solutions that were approved by other research teams and that were already part of the derivatization method were tested. This helped to avoid introducing additional ions through matrix effects.

Another challenge was adjusting the concentration of the extract solution in order to achieve a high extraction efficiency at low matrix effects. First, the spiked Eurosoil samples with KOH and tetraborate buffer at a concentration of 0.1 M KOH and 5% tetraborate buffer was tested. The measurements showed that the recovery with KOH extraction (17-32%) was lower than for samples extracted with tetraborate buffer (98-108%), both series with RSD < 10%. The results for the analyzed samples extracted with tetraborate showed higher chromatography efficiency due to lower matrix effects.

In the next step, different concentrations of the tetraborate buffer to obtain the highest extraction efficiency for the investigated substances were tested. 40 mM tetraborate, which is also part of the derivatization method, and a much higher concentration of 5 % of the same buffer, were used. The measurements showed similar extraction yields. With both extraction medias the mean recovery was ~ 120 % (RSD <10%). However, due to lower matrix effects and the compatibility with the derivatization procedure, the lower concentration was chosen for all further experiments.

The strategy of developing an extraction procedure for glyphosate and AMPA mainly focused on the adsorption of the compounds to iron oxides of the soil matrix. The most common iron oxide in matrixes of soils developed under moderate climatic conditions is goethite, which is mainly responsible for anion exchange capacity in soils. The first step in extracting glyphosate and other ionic metabolites from soil matrices was to increase the pH of the extractable soil suspension to reach pH values of the point of zero charge (PZC) of goethite. This could be done with solutions of potassium hydroxide, sodium hydroxide or sodium tetraborate in the range of pH 8.5. However, increasing the strength of the alkaline extraction solution causes a coelution of the dissolved organic carbons (humic compounds of the soil matrix) which disturbs the chromatographic performance of the method.

It was not only a question of recovery, but also of LC/MS performance and the capacity of the SPE system to take a 40 mM sodium tetraborate solution to extract glyphosate compounds from soil matrix. This extracting solution is sufficient to release glyphosate from the soil matrix with a minimum of co-extraction of humic compounds and a minimum of inorganic salts in the sample. The maximum sodium ion concentration for the electron ion spray of the mass spectrometry system is approximately 100 mM.

To prove the suitability of the chosen extraction solution, it was tested with three different spike concentrations: 100, 200 and 500 µgkg^{-1} glyphosate. The recovery for glyphosate was between 77 and 91% (RSD <10%), and for AMPA 61-94% (RSD <2 %) with five replicates. The lower recovery was found for the samples with the lower concentration of glyphosate and AMPA (100 µg kg^{-1}) because it was much nearer to the limit of detection (LOD) for these substances.

Based on all these results, a 40 mM tetraborate buffer was used to extract glyphosate and AMPA because the results had shown high recovery and acceptable repeatability. The next problem, beyond identifying a suitable extraction solution and concentration, was to optimize the clean-up process with SPE cartridges to avoid the excessive extraction of humic substances that would interfere with measurements using mass spectrometry, Hennion (1999).

In Fig. 9 an example chromatogram of determination of the glyphosate from a Kirchberg soil sample with and without SPE clean-up is shown.

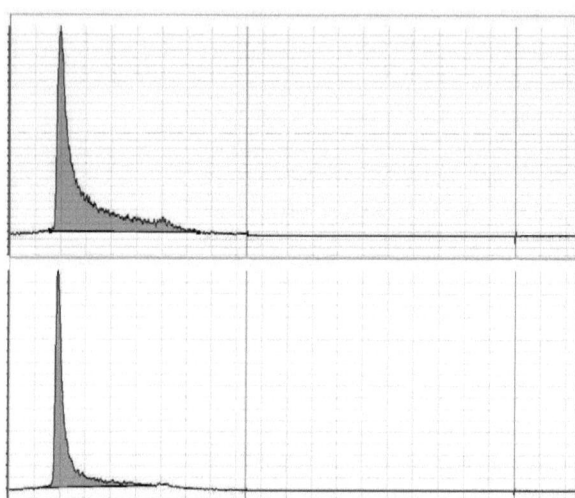

Figure 9: Acquisition resulting in the typical total ion current plot (TIC) of glyphosate without (above) and with SPE clean-up (below) from a soil sample of Kirchberg.

Without the clean-up procedure the glyphosate peak has a long tail. The humic substances interfere with the chromatographic conditions of the separation of glyphosate and AMPA on the analytic column of Zorbax Eclipse C8 XDB from Agilent Technologies Inc. (Santa Clara, CA, USA). The SPE cartridges (strata-X) trap the co-eluated humic substances by the extraction procedure.

One of the main characteristics of glyphosate is the complexation with metal ions, (Morillo et al., 2000; Gimsing et al., 2004; Schnurer et al., 2006; Ghanem et al., 2007). In order to prevent potential binding of the herbicide and false measurements, EDTA was added to guard the derivatization process. To improve the sensitivity in the measurements of glyphosate and AMPA in the soil samples, and to minimize possible mistakes in the derivatization procedure, the ultratraces analytic method for water analysis by Hanke et al., (2008) was used, after adaptation for soil analysis.

For better mass spectrometry conditions we used 9-fluorenylmethylchloroformate (FMOC) for derivatization of both analyzed substances (Hogendoorn et al., 1996; Vreeken et al., 1998; Sancho et al., 1996). The derivatization procedure is highly pH sensitive and, in the case of inappropriate pH, derivatization of glyphosate and AMPA molecules can fail (Ibáñez et al., 2006).

Ibáñez et al., (2006) showed a similar solution for recovery improvement based on modifying the derivatization procedure. In order to control the derivatization process, isotope-labeled standards for both analyzed substances were used. Matrix complexity of soil samples could effect a co-extraction of the humic substances, which could interfere with the quantification of the analyzed substances (Ibáñez et al., 2005).

In Fig. 10 and 11 masstransitions, retention time and mass spectra for glyphosate and AMPA, as well for labeled internal standards, are shown. The contaminant was identified through two identification points, e.g. one precursor and one product ion for each of the both substances (see Tab. 5). The retention time for glyphosate was 4.9 min, for AMPA 5.8 min.

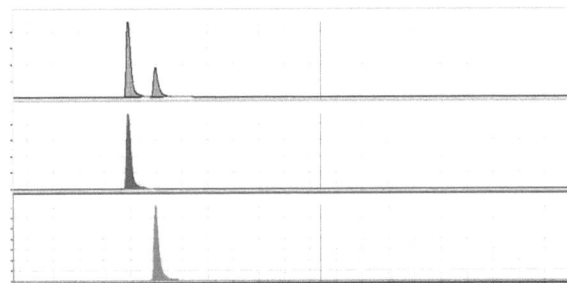

Figure 10: Masstransitions for FMOC-glyphosate (390→168) and FMOC-AMPA (332→110), TIC, MRM (MS multi reaction mode) and retention time (RT min) from a 500 µg L^{-1} Mix-Standard.

Figure 11: MRM Production mass spectra from FMOC-glyphosate (C13, N15), FMOC-glyphosate, FMOC-AMPA (C13, N15), FMOC-AMPA. FMOC-glyphosate (C13, N15) and FMOC-AMPA (C13, N15) were used as internal standards (ISTD).

3.2.2 Results of the validation study

The linearity of the method calibration was evaluated analyzing four standard solutions for glyphosate in the range 50-500 µg L^{-1} with r^2 = 0.9972 (n = 4) and five for the range 100-1500 µg L^{-1} with r^2 = 0.9986 (n = 5).
For AMPA, the calibration curve was determined by analyzing five standard solutions in the range 50-750 µg L^{-1} with r^2 = 0.9983 (n = 5).

The internal precision (repeatability, expressed as relative standard deviation (STD in %) for glyphosate measurements for five replicates of 100 µg L^{-1} was 7.89 (RSD = 9.27%) with a recovery mean value of 85.1%. For AMPA measurements STD was 1.10 (RSD = 1.81%) with recovery mean value of 69.9%. The linear calibration curves were obtained for both substances investigated: for glyphosate in the range 5-1000 µg kg^{-1} and for AMPA 5-1000 µg kg^{-1}. The correlation coefficient (r^2) was higher than 0.995 for all measurements. The RSD was below 10%. The recoveries for glyphosate were in the tolerable range of 70-120 %. The recoveries for AMPA were above 200 µg.kg^{-1} confirming the satisfactory results of the measurements.

The intraday precision for glyphosate was 3.4% and the day to day precision 7.1%. The intraday precision for AMPA was 5.3%, and the day to day precision 8.4%.

The determination of the limit of quantification (LOQ) (ISO 3534-1, Geneve) was numeric, STD of counts from 3-3.5 min, 10 times STD divided by the K-value. The K-value is the peak height in counts divided by the nominal concentration of the standard.

The limit of detection (LOD) was obtained when the signal was three times the STD divided by K-value.

Tab. 7 presents recovery, LOD and LOQ of the analyzed compounds of the three soils investigated.

Table 7: Recovery, LOD and LOQ of the analyzed compounds of the three soils investigated.

	Glyphosate			AMPA		
	Kirchberg	Pixendorf	Phyra	Kirchberg	Pixendorf	Phyra
Recovery (%)	93.5	95.7	69.9	92.4	98.1	69.9
RSD (%)	<2	<2	<4	<2	<2	<4
LOD ($\mu g\ kg^{-1}$)	6.8	4.3	13.8	26.7	25.2	87.2
RSD (%)	<10	<10	<7	<10	<10	<9
LOQ ($\mu g\ kg^{-1}$)	22.7	14.4	46.5	88.9	84.0	120.3
RSD (%)	<5	<6	<6	<2	<5	<8

3.2.3 Results of the quality control

The quantification of the analyzed samples was considered satisfactory if the recoveries were in the range of 70-120%. The procedural precision of the measured concentrations expressed as relative standard deviation was accepted in a range of ±20%. In order to obtain high-quality results, control standards and blank values were included and measured with the sample series.

In order to test whether the extraction method is applicable for different soils, three different soils with major contrasting physico-chemico-mineralogical characteristics from the experimental fields of Kirchberg, Pyhra and Pixendorf were analyzed (see Tab. 8-13) where Roundup Max was applied according to standard agricultural practice in Austria (1800 g glyphosate ha^{-1} corresponding to 180 mg glyphosate m^{-2}). The glyphosate and AMPA contents in the three analyzed soils of Kirchberg, Phyra and Pixendorf are shown in Fig. 12.

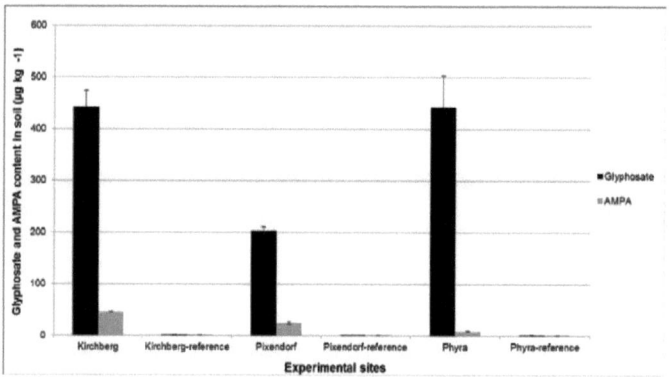

Figure 12: Glyphosate and AMPA contents in the investigated soils at 0-2 cm soil depth after conventional Round Up Max application. Values as average from the 3 field replications. The reference values show the glyphosate contents before application (e.g. the residual traces, next to zero, from the previous application one year before) at each site.

The results presented in Fig. 12 show that the newly developed extraction method clearly distinguishes the different contents of glyphosate and AMPA in different soils according to their physico-chemical adsorption properties.

Based on previous studies, it was expected that glyphosate and AMPA would be more strongly adsorbed in soils with a higher iron-oxide (Fed) content (Mentler et al., 2007; Pessagno et al., 2008). The results of this work show that both the Kirchberg-cambisol, and the Phyra-stagnosol, with a higher pedogenic iron-oxide content, respectively 16000 and 10000 mg Fed . kg^{-1} soil, (see Tab. 11), adsorbed a distinctly higher quantity of glyphosate and AMPA than the Pixendorf –chernozem which had a distinctly lower iron-oxide content, (7900 mg Fed . kg^{-1} soil, see Tab. 11), and a Kd-value about 10 times lower than that from Kirchberg (Klik et al., 2010).

Moreover, the weakly weathered Pixendorf-chernozem consequently shows a low content of amorphous (Feo) iron-oxides (973 mg Feo . kg^{-1} soil) with respect to the more highly weathered Kirchberg- cambisol (3402 mg Feo . kg^{-1} soil) and Pyhra-stagnosol (3279 mg Feo . kg^{-1} soil), see Tab. 11. Higher content of pedogenic iron-oxides (Fed) (Barja and dos Santos Afonso, 1998; Barja et al., 2001; Zhou et al., 2004; Gimsing et al., 2004; Morillo et al., 2000) and even higher contents of amorphous iron-oxides (Feo) (Piccolo et al., 1994; Gerritse et al., 1996), lead to higher sorption of glyphosate and AMPA, probably due to a larger and more reactive surface area of amorphous iron-oxides.

Thus, iron-oxides in general seem to be a key parameter for glyphosate and AMPA adsorption in soils. This study confirmed this: the analysis showed lower contents of Fed and Feo for the Pixendorf-chernozem, with consequently lower adsorption of glyphosate and AMPA compared with the Kirchberg-cambisol and the Pyhra-stagnosol (see also the results of Eberbach, 1997).

The Phyra and Kirchberg soils are acidic, with a pH of 5.1 and 5.7 respectively, whereas the Pixendorf –chernozem is neutral to alkaline (pH 7.3), (see Tab. 10). McConnell and Hossner (1985) showed how the pH-value of soils strongly influences the adsorption behaviour of glyphosate and AMPA. These circumstances also explain the better adsorption of the two investigated substances at Kirchberg and Phyra.

The higher content of clay minerals in the Phyra-stagnosol (see Tab. 13), could also influence the more pronounced adsorption of both substances investigated, as well as the acidic pH-value, as in some previous studies (e.g. da Cruz et al., 2007; Morillo et al., 1997; Piccolo et al., 1994; and Damonte et al., 2007) also stressed the importance of expandable clay minerals like smectite and vermiculite for the adsorption of the investigated herbicide. This study would confirm this influence at Phyra, where the amount of expandable clay minerals smectite and vermiculite together was >50% (results not shown), but not in Kirchberg and Pixendorf, where the smectite and vermiculite contents were low.

The method shows no interference with the amount of iron oxides in the soil matrix, but high concentrations of clay minerals could have an influence on the recovery and RSD (see Tab. 8). This may be related to dispersion problems of clay minerals in heavy soils (e.g. Phyra- stagnosol).

3.3 Adsorption of glyphosate and AMPA in agricultural soils

The results in the literature show that iron oxides play an important role in the soil retention of glyphosate. From the investigated iron oxides (ferrihydrite, hematite and goethite) ferrihydrite had the highest impact on the adsorption process of glyphosate (Mentler et al., 2007).

The variation of the KD-values in different soils is significant (Mentler et al., 2007) and seems to depend mainly on the iron oxide content, see Tab. 14.

Table 14: Soil properties and KD-values for glyphosate for different soils and silica sand based on literature data (after Mentler et al., 2007).

	KD-value [l/kg]	pH value	Clay [w/w%]	Corg [w/w.%]	Fe [w/w.%]	Location
Mentler et al., 2007	467 - 519	4.5	2.7	0.8	3.2	Wienerwald
Mentler et al., 2007	13.8 – 29.3	5.8	17.2	3.45	2.2	Phyra
Mentler et al., 2007	188 - 299	5.2	18.8	6.7	2.1	Eurosoil 7
Mentler et al., 2007	1.5 – 2.9	6.4	<0.1	<0.01	< 0.01	Silica sand
Sorensen et al., 2006	271 - 385	4.3-5.6	2-4	0.1-4.9	0.01-0.05	Fladerne Beak
Mamy et al., 2005	13.2 - 31.1	8.2-8.5	8.8-9.5	1.3-2	0.16-0.19	Chalons

Site Pixendorf

The results of the investigations at the Pixendorf-chernozem are shown in Fig. 13-14.

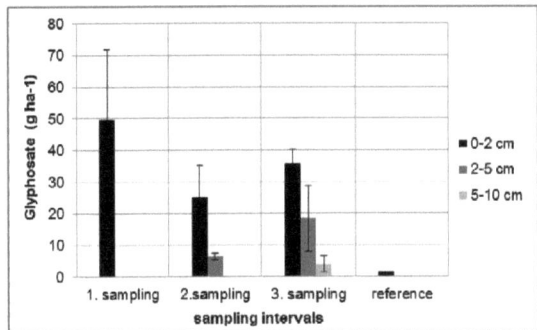

Figure 13: Pixendorf-chernozem: glyphosate contents in soil at different time intervals and soil depths. (1. Sampling = immediately after application; 2. Sampling = 3 days after application; 3. Sampling = 10 days after application; reference = residues before application).

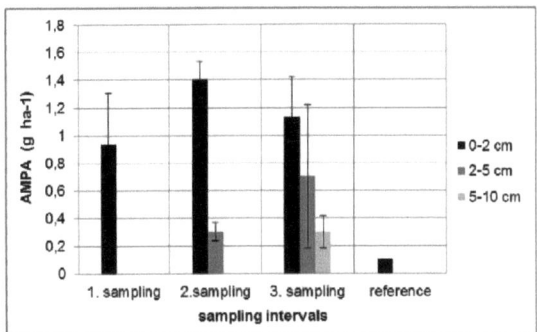

Figure 14: Pixendorf-chernozem: AMPA contents in soil at different time intervals and soil depths. (1. Sampling = immediately after application; 2. Sampling = 3 days after application; 3. Sampling = 10 days after application; reference = residues before application).

At the first sampling after field application of Round Up Max about 30% of the applied glyphosate amount was detected in the upper 0-2 cm. The main part of the herbicide adheres at the green plant cover and at first does not enter the soil surface.

After 3 days the glyphosate content decreased in the topsoil and was transported and adsorbed in the next horizon (2-5 cm) with concomitant increase of the AMPA content. After 10 days the glyphosate content was higher than immediately after application. For this behavior following hypothesis can be possible: a) plant adsorbed glyphosate is released to the topsoil after partly decomposition of the weeds; b) During the time between second and third soil sampling about 10 mm precipitation fell down, this may have washed glyphosate from plant leaves out. The increase of AMPA 3 days after application of Round Up Max shows the very quick degradation of glyphosate to AMPA. The fact that AMPA was detected in soil samples collected immediately after application of Round Up Max is explainable through the time between sampling in the field and freezing in the laboratory (about 2 hours). It can also not be excluded that a degradation from glyphosate to AMPA could probably take place already in the Round Up package, since the farmers surely do not use every time new Round Up packages.

Site Pyhra

The results of the investigations at the Pyhra-stagnosol are shown in Fig. 15-16.

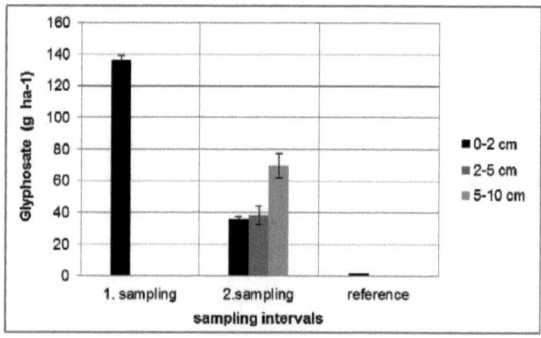

Figure 15: Pyhra-stagnosol: glyphosate contents in soil at different time intervals and soil depths. (1. Sampling = immediately after application; 2. Sampling = 28 days after application; reference = residues before application).

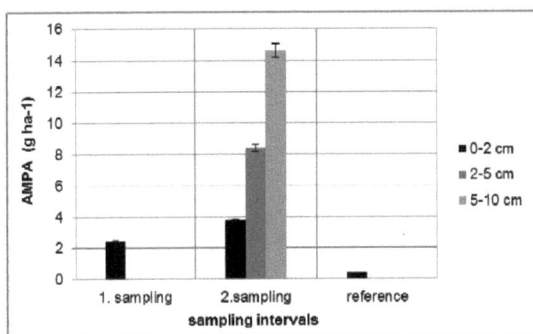

Figure 16: Pyhra-stagnosol: AMPA contents in soil at different time intervals and soil depths. (1. Sampling = immediately after application; 2. Sampling = 28 days after application; reference = residues before application).

Most of the applied glyphosate was transported and adsorbed in deeper horizons after 28 days. The reference glyphosate and AMPA values in Fig. 15 and 16 refer the amount of both substances before application, e.g. the residues oft he previous application (normally 2 years before). That means that in the Pyhra-stagnosol glyphosate is transported downwards within 2 years and probably bound to deeper soil layers.

Site Kirchberg

The results of the investigations at the Kirchberg-cambisol are shown in Fig. 17-18.

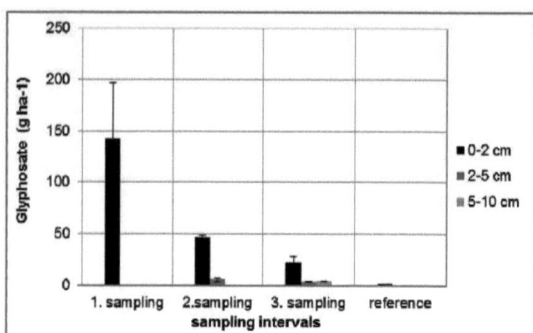

Figure 17: Kirchberg-cambisol: glyphosate contents in soil at different time intervals and soil depths. (1. Sampling = immediately after application; 2. Sampling = 3 days after application; 3. Sampling = 12 days after application; reference = residues before application).

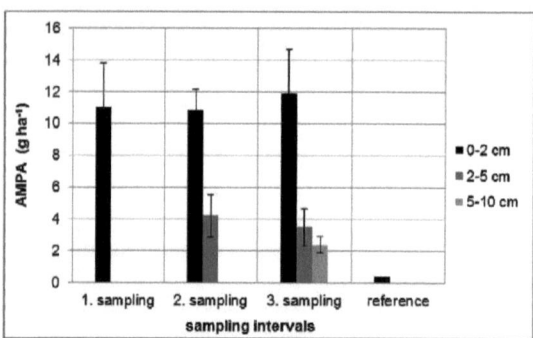

Figure 18: Kirchberg-cambisol: AMPA contents in soil at different time intervals and soil depths. (1. Sampling = immediately after application; 2. Sampling = 3 days after application; 3. Sampling = 12 days after application; reference = residues before application).

The Kirchberg-cambisol features the best potential adsorption capacity for glyphosate with about 16.000 mg Fe$_d$.kg^{-1} soil and about 3.500 mg Fe$_o$.kg^{-1} soil, see Tab. 4, but previous studies with rain simulation experiments (Rampazzo Todorovic *et al.*, 2010) showed that that site can be strongly influenced by erosion processes if the infiltration rate for rainfall is reduced by soil crusting.

This is the reason why glyphosate strongly decrease in the upper soil horizons but does not accumulate in deeper horizon. A considerable amount of the applied glyphosate is transported downslope with run-off (Rampazzo Todorovic et al., 2010). Moreover, the degradation from glyphosate to its metabolite AMPA is visible by the increase of AMPA with time.

3.4 Rainfall simulation experimental results

Pixendorf

Fig. 19 and 20 show the amount of run-off after rain simulation experiments at Pixendorf.

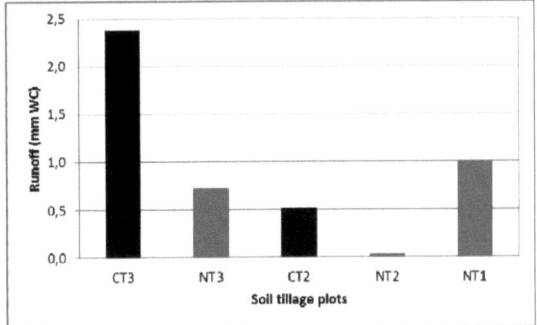

Figure 19: Pixendorf-chernozem: Total run-off of the CT- and NT- plots (CT = Conventional Tillage, NT = No Tillage) in the 3 field replications (Plot CT1 could not be exercised).

At Pixendorf as well as at Kirchberg the CT-plots produced the highest run-off amounts because of the lower vegetation cover see also Fig. 26. The amount of run-off at Pixendorf was 10 times lower than at Kirchberg because of its very favorable crumby structure with a high infiltration rate during the rainfall simulation, whereas the soil surface of Kirchberg was compacted and crusty. The pronounced different amounts of run-off between the 3 field replications show the high inhomogeneity of the soil conditions.

Fig. 20 shows the time dependent glyphosate concentrations in run-off-fractions at time intervals of respectively 15 min-at Pixendorf. As it was expected the first fractions (respectively on the left) showed the highest concentrations in both variables CT and NT which got lower with time. The CT-plots showed distinguished higher concentrations than the NT-plots because of the lower vegetation cover.

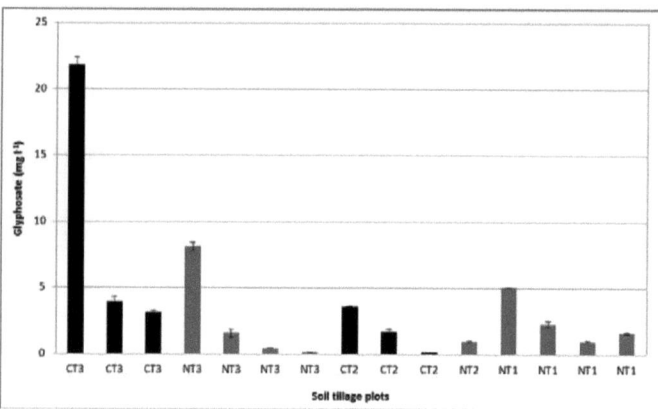

Figure 20: Pixendorf-chernozem: Glyphosate concentrations in water run-off-fractions at different time intervals in the 3 field replications (1, 2, 3) plots (CT = Conventional Tillage, NT = No Tillage). (Plot CT1 could not be exercised).

In concomitance with the concentration, the total amount of glyphosate washed out of the plots by run-off at Pixendorf was much higher in the CT-plots, see Fig.21, but still much lower than in Kirchberg, see Fig.28.

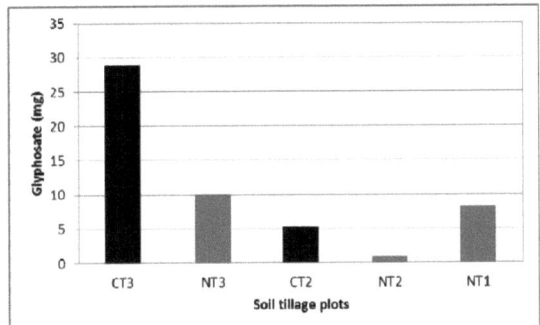

Figure 21: Pixendorf-chernozem: Total amounts of glyphosate in water run-off at the 3 field replications (1, 2, 3) plots (CT = Conventional Tillage, NT = No Tillage). (Plot CT1 could not be exercised).

The content of glyphosate in the solid phase of run-off at Pixendorf is showed in Fig. 22.

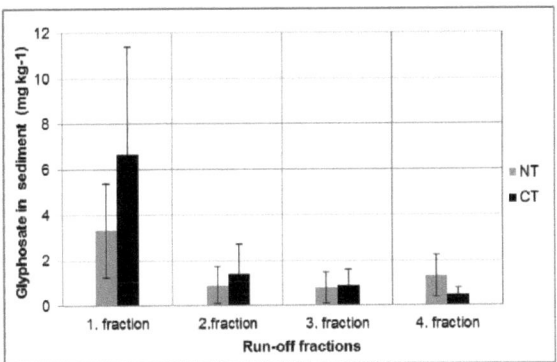

Figure 22: Pixendorf-chernozem: Glyphosate contents in the solid phase of run-off after 4 time intervals (= fractions as average of the 3 field replications).

The first collected fraction of run-off contains the highest amounts of glyohosate which then generally decreases in the following fractions. The CT-plots shows higher amounts of solid phase than the NT-plots because of the lower vegetation cover. Analogue is the distribution of AMPA in the sediment, see Fig. 23.

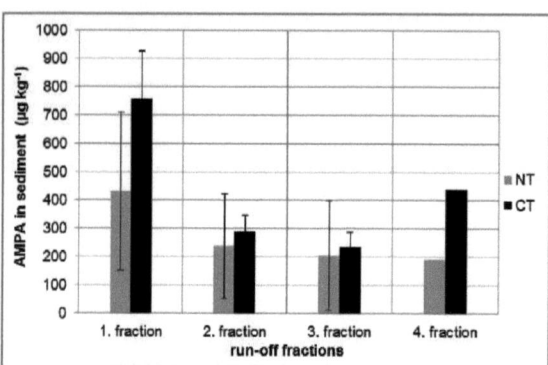

Figure 23: Pixendorf-chernozem: AMPA contents in the solid phase of run-off after 4 time intervals (= fractions, as average of the 3 field replications).

Fig. 24-25 show the content of glyphosate and AMPA adsorbed by the soil immediately after the rain simulation experiments at Pixendorf.

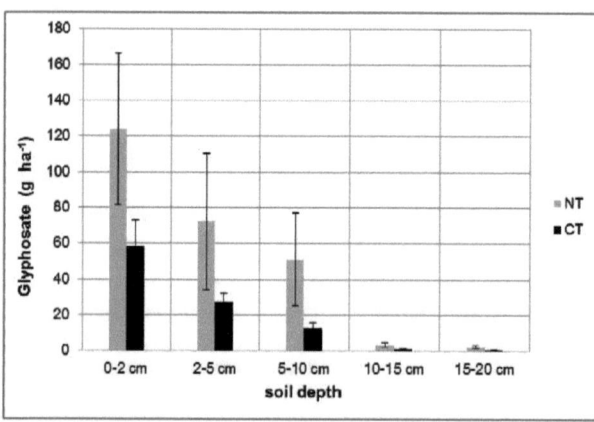

Figure 24: Pixendorf-chernozem: Glyphosate contents in the soil within the rain simulation plots (average value from the 3 field replications).

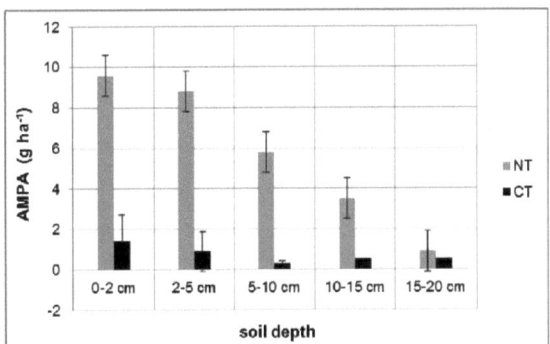

Figure 25: Pixendorf-chernozem: AMPA contents in the soil within the rain simulation plots (average value from the 3 field replications).

Finally, Fig. 24-25 shows a clear depth function of the adsorption of glyphosate and AMPA through the soil immediately after Round Up Max application and rainfall simulation at Pixendorf. The glyphosate and AMPA contents clearly decreased with soil depth. As a previous study by Mamy et al., (2005) showed, this could be explained because the two compounds are very quickly adsorbed by the soil compounds, probably also depending on the physical soil conditions and water flow during rainfall. In fact, the soil had a favourable crumby structure, with no cracks, no preferential flow, and optimal conditions for water retention in the upper soil layers at the moment of the rainfall simulation experiment, so that more than 50% of the adsorbed glyphosate was retained in the first 5 cm of the soil.

The fact that AMPA could already be detected 1 h after the Round Up Max application underlines the quick glyphosate degradation in soil (Mamy et al., 2005).

A tentative total balance of the glyphosate distribution after the rain simulation experiment at Pixendorf is shown in Tab. 15:

Table 15: Pixendorf-chernozem: Distribution of glyphosate (measured) in % of the applied quantity into the 2m x 2m rain simulation plots (720 mg glyphosate = 100%) P.S. The plot repet.1-CT could not be exercised.

	run-off-liquid %	run-off-solid %	soil %	remain %
Repetition 1				
NT (No Tillage)	1.14	0.009	74.1	24.7
CT (Conv. Tillage)	-	-	-	-
Repetition 2				
NT (No Tillage	0.13	0.00012	64.5	35.4
CT (Conv. Tillage)	0.73	0.0036	47.3	51.9
Repetition 3				
NT (No Tillage)	1.40	0.006	50.0	48.6
CT (Conv. Tillage)	4.00	0.0081	34.4	61.6
Mean value NT	0.89	0.005	62.94	36.2
Mean value CT	2.36	0.0043	40.9	56.7

The 3 field repetitions were relatively inomogenous. The CT-plots showed higher glyphosate % contents in the liquid phase of run-off than the NT-plots. The glyphosate and AMPA adsorption in soil was much higher in Pixendorf as compared to Kirchberg, (see Tab. 16) because of the high infiltrability of the soil. The solid phase of run-off contained negligible amounts of glyphosate. The rate of detection in % of the applied glyphosate amounts at Pixendorf was relatively high, and much higher than at Kirchberg.

Kirchberg

Fig. 26 and 27 show the amount of run-off after rain simulation experiments at Kirchberg.

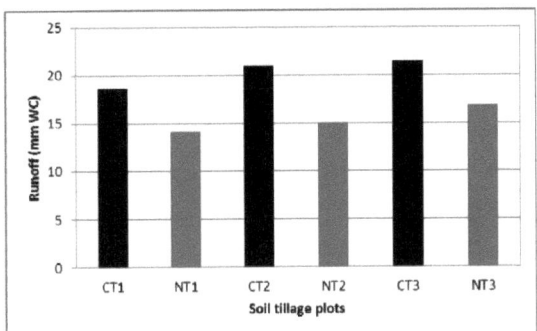

Figure 26: Kirchberg-cambisol: Total run-off of the CT- and NT plots (CT = Conventional Tillage, NT = No Tillage) in the 3 field replications.

Also at Kirchberg the CT-plots produced considerable higher run-off amounts than the NT-plots because of the lower vegetation cover. Moreover, the amount of run-off in Kirchberg was 10 times higher than at Pixendorf (see Fig.19) because during the rainfall simulation the soil surface of Kirchberg was compacted and crusty, whereas Pixendorf had a very favorable crumby structure with a high infiltration rate.

The glyphosate concentration in run-off at Kirchberg is shown in Fig. 27.

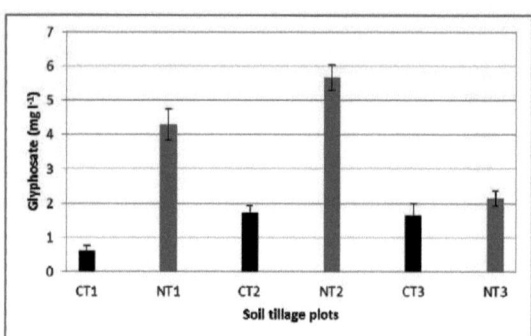

Figure 27: Kirchberg-cambisol: Average glyphosate concentrations in water run-off at the 3 field replications (1, 2, 3) plots (CT = Conventional Tillage, NT = No Tillage).

In this case the run-off of the NT-plots had a much higher concentration of glyphosate than the CT-plots. The reason for this is the nearly 100% plant cover of the NT-plots, where most of the applied glyphosate adheres to the photosynthetic active plant organs (still and leaves), get washed out through the rainfall und hardly infiltrates the soil surface. Moreover, the concentrations in the 3 field repetitions were different because of the inhomogeneity of the field regarding soil surface structure, vegetation cover and micro relief.

Fig. 28 shows the total glyphosate amounts in the run-off of the erosion plots, where controversial to the site Pixendorf, see Fig. 20, the NT-plots had distinguished higher contents.

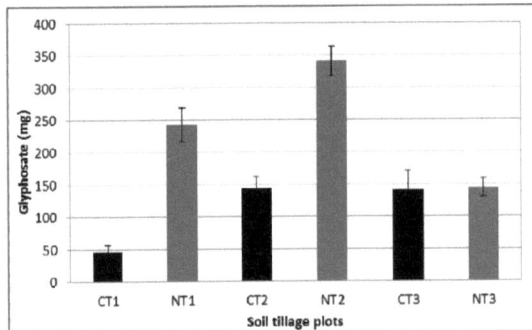

Figure 28: Kirchberg-cambisol: Total amounts of glyphosate in water run-off at the 3 field replications (1, 2, 3) plots (CT = Conventional Tillage, NT = No Tillage).

The content of glyphosate and AMPA in the solid phase of run-off at Kirchberg is showed in Fig. 29 and 30 respectively.

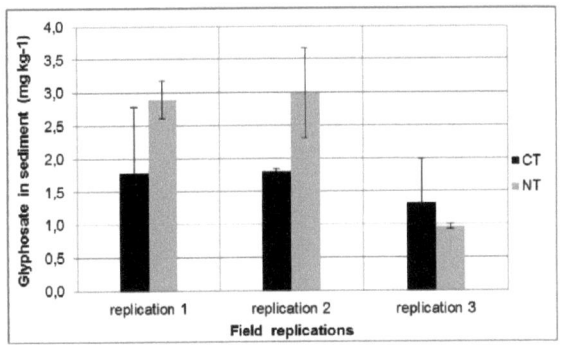

Figure 29: Kirchberg-cambisol: Glyphosate contents in the solid phase of run-off at the 3 field replications (1, 2, 3) plots (CT = Conventional Tillage, NT = No Tillage).

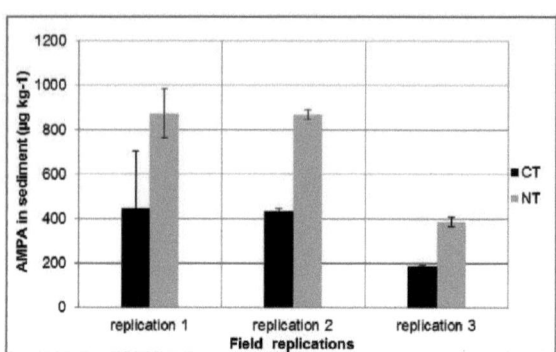

Figure 30: Kirchberg-cambisol: AMPA contents in the solid phase of run-off at the 3 field replications (1, 2, 3) plots (CT = Conventional Tillage, NT = No Tillage).

The concentrations of glyphosate and AMPA in run-off at Kirchberg are distributed similar to the corresponding aqueous fractions of run-off; they are mostly higher in the NT-plots than in the CT-plots.

Fig. 31-32 show the content of glyphosate and AMPA adsorbed by the soil immediately after the rain simulation experiments at Kirchberg.

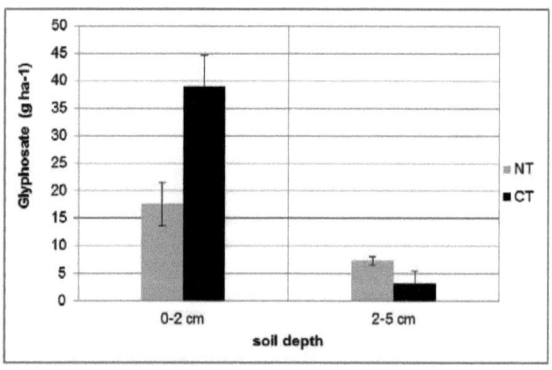

Figure 31: Kirchberg-cambisol: Glyphosate contents in the soil within the rain simulation plots (average value from the 3 field replications).

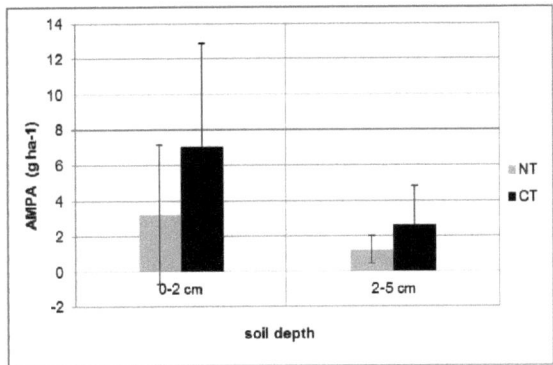

Figure 32: Kirchberg-cambisol: AMPA contents in the soil within the rain simulation plots (average value from the 3 field replications).

Immediately after the rain simulation experiment a very clear distribution appears: glyphosate and AMPA are first adsorbed in the upper 0-2 cm of the soil and only a few amount reaches the next soil depth of 2- 5 cm. In general the NT-plots show a distinguished lower content on glyphosate and AMPA as compared to the CT-plots. This is explained by the respectively higher glyphosate contents in run-off of NT, see Fig. 27-28.

A tentative total balance of glyphosate for the rain simulation experiment in Kirchberg is shown in Tab. 16:

Table 16: Distribution of glyphosate in % of the applied quantity into the 2m x 2m rain simulation plot (720 mg glyphosate = 100%) of Kirchberg.

	run-off-liquid %	run-off-solid%	soil %	remain %
Repetition 1				
NT (No Tillage)	33.7	0.025	13.0	53
CT (Conv. Tillage)	6.3	0.07	0.1	93
Repetition 2				
NT (No Tillage	47.2	0.019	6.4	46.4
CT (Conv. Tillage)	19.9	0.06	17.8	62.2
Repetition 3				
NT (No Tillage)	20.0	0.016	15.3	64.7
CT (Conv. Tillage)	19.6	0.052	33.1	47.2
Mean value NT	33.6	0.02	11.6	54.7
Mean value CT	15.3	0.06	17.0	67.5

The 3 field repetitions were also in Kirchberg relatively ihnomogenous. The NT-plots showed higher glyphosate % contents in the aqueous run-off than the CT-plots. The glyphosate adsorption in the soil was relatively low. The solid phase of run-off contented also in Kirchberg negligible amounts of glyphosate. The CT-plot of repetition 1 was extremely crusty because only about 7% (solid + liquid run-off) of the applied glyphosate could be measured. Averagely (with exception of rep.1-CT) about 50 % of the applied glyphosate could be measured, respectively 50 % not found (remain).

Soil loss trough erosion:

The soil losses through erosion at Pixendorf and Kirchberg is shown in Fig. 33-34.

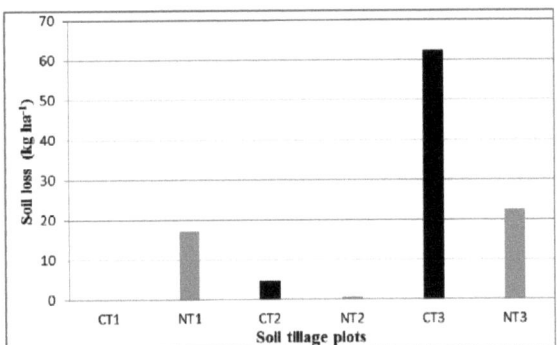

Figure 33: Pixendorf–chernozem: soil loss after the rain simulation experiments in the 3 field replications (1, 2, 3, plot CT1 could not be experimented) at different tillage plots (CT = Conventional Tillage, NT = No Tillage).

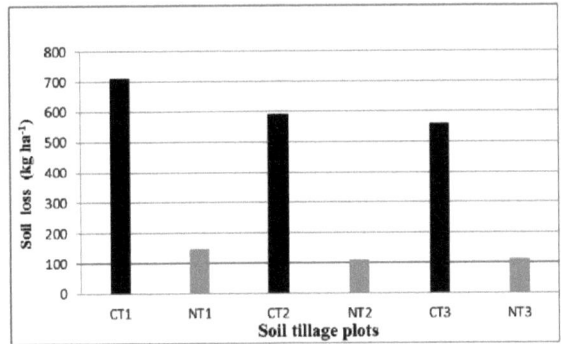

Figure 34: Kirchberg–cambisol: soil loss after the rain simulation experiments in the 3 field replications (1, 2, 3) at different tillage plots (CT = Conventional Tillage, NT = No Tillage).

At both sites the soil loss from the CT-plots was (at Kirchberg much higher) higher than from the NT-plots. This was due to the different vegetation cover degree between CT- and NT-plots before the simulation experiment, see Fig. 5-6. The soil loss at Kirchberg was 10 times higher than at Pixendorf. The reason for this is that the two experimented sites had a completely different soil structure and surface the

Pixendorf-chernozem had a very friable, crumby, permeable stucture after the wheat yield. The Kirchberg-cambisol stood right after the corn yield, the soil surface was crusty and less permeable.

3.5 Dissipation of glyphosate and AMPA through natural erosion and leaching processes

Both rain simulation experiments were intentionally carried out as „worst case scenario", e.g. a heavy erosive rainfall falls immediately after the Round Up Max application at field conditions.
The site Pixendorf and surrounding is generally known as a location with high erosion risk because of the high silt amount (> weight 60%) and especially with corn crop, where deep gully erosion takes place, see Fig. 35.

Figure 35: Pixendorf–chernozem: rill erosion (details) at the experimental field during corn crop.

The erosion rills discharge downslope to an artificial run-off retention basin at footslope of the experimental field. This basin can run over and flow downwards on different paths and is collected through further toeslope retention basins, see Fig. 36-378.

Figure 36: First retention basin at the footslope of the experimental field at Pixendorf.

Figure 37: Second retention basin toeslope at the toeslope of the experimental field at Pixendorf.

Water samples from both retention basins were collected (2009 and 2010) and analysed for glyphosate, see Fig. 38.

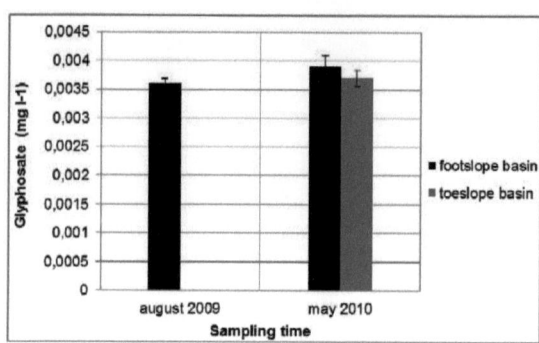

Figure 38: Glyphosate contents in natural run-off retention basins.

Moreover, soil percolation water samples at footslope of the experimental field were collected at two depths from previously installed stations and analysed for glyphosate and AMPA. Since the rainfall simulation experiment was conducted topslope whereas the percolation water samples where collected at footslope (100 m distance) at the same time, it seems unlikely that the measured amounts of glyphosate and AMPA were influenced by the rain simulation, but are probably residual amounts from previous field application, confirming the possibility of dissipation through natural processes.

The results show that small amounts of glyphosate and its metabolite can dissipate through soil percolation, mainly depending on the physico-chemical adsorption as well structural properties of soils, see Fig. 39-40.

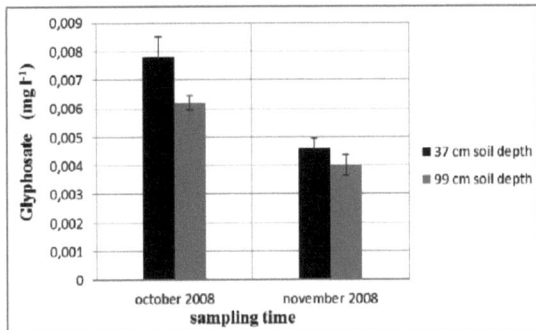

Figure 39: Glyphosate contents in percolation water at 2 different perionds and soil depths.

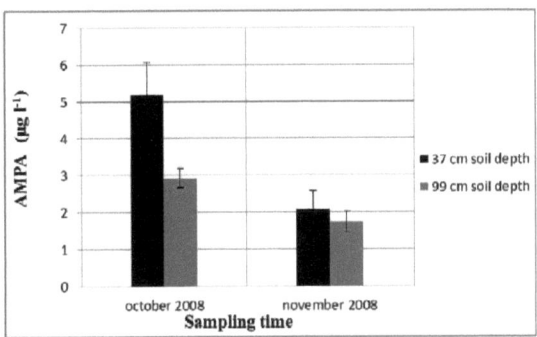

Figure 40: AMPA contents in percolation water at 2 different perionds and soil depths.

In the autumn of the the rain simulation experiment (october 2008) the Pixendorf-chernozem had a very favourable permeable crumby strcture because a green vegetation cover was growed during the summertime. The rainfall simulation experiment showed that the amounts of run-off and so the dispersion of glyphosate and AMPA were not so bad as expected (the glyphosate transport with run-off was averagely 8% of the applied amount in the NT-plots and 16% in the CT-plots respectively) and most of the glyphosate could be adsorbed by the soil, see Tab. 15.)

This shows the importance of protection practices like intermediary crops and so amelioration of soil structure and soil infiltration for rain.

4 Conclusions

Development of a new method for the determination of glyphosate and AMPA with a HPLC-MS/MS method.

The success of the new extraction method was confirmed by measuring the spiked soil samples to control recovery and the detection limits. This was followed by applying the same procedure to different soil types with contrasting characteristics from the experimental fields after glyphosate application. The method developed covers matrix effects of the most representative agricultural soil types of Austria but can be applied for glyphosate and AMPA investigations on any kind of mineral soils. The method, which could precisely and reliably identify glyphosate and AMPA in soils even at low concentrations, is sensitive to co-extractable humic substances. For the investigation of organic soils or organic layers the SPE-extraction (clean-up) should be adapted (the amount of solid phase in the cartridge should be enhanced to increase the capacity of the clean-up procedure).

Moreover, each step of the soil sampling campaign, preparation and measurement was chosen and adopted for reliable and precise laboratory practice. In our measurement procedure with LC-MS/MS, a negative mode was used and, with FMOC, derivatized glyphosate and AMPA ions were identified based on the precise determination of the ion mass. The LOQ of both substances (13.8-22.7 µg kg^{-1} for glyphosate (RSD <10%) and 84.0-88.9 µg kg^1 (RSD <10%) for AMPA) in all three soil types demonstrates the sensitivity of the method, which enables measurements of both substances in different soil matrixes. The method allows to determine the analyte (glyphosate) and the main metabolite (AMPA). From this, a calculation of the degradation ratio of glyphosate is possible, except an input of further detergents (with the same degradation pathway to AMPA like glyphosate) to the investigated soil (e.g. through additives for pesticide application, irrigation with AMPA-contaminated water, etc.) took place.

The tandem mass spectrometry method with FMOC derivatization has the highest selectivity for glyphosate and AMPA. The post column derivatization method with florescence detection has a high vulnerability because the ion chromatography is very sensitive to humic matrixes and has a lower selectivity for glyphosate and AMPA. A tandem mass spectrometry method without FMOC derivatization is much more matrix sensitive because of the low mass of glyphosate.

Based on soil analysis of the three soils investigated from Pixendorf, Phyra and Kirchberg, the influence of higher contents of iron-oxides (Fed and Feo), clay content and acidic pH on a more pronounced adsorption of glyphosate and AMPA in soil is demonstrated.

Since the extraction is buffered with 40 mM tetraborate buffer at pH 8.5, e.g. close to the zero point of charge (ZPC) of Fe-oxides, and humic substances with variable positive electric charges are co-extracted, the method focuses mainly to the glyphosate adsorption by Fe-oxides and organic matter. Adsorption by different clay minerals is theoretically unlikely because of their negative net electric charge but should be investigated by further studies.

Adsorption of glyphosate and AMPA in agricultural soils.

The three investigated soils, the most representative agricultural soils in Austria, differ through major contrasting chemico-mineralogical characteristics, according to their genetical formation.

The Pixendorf-chernozem shows the development from loess with a typical high silt content. The Pyhra-stagnosol has a heterogeneous texture, typical for a loamy soil and the high amount of silt and clay explains the water stagnation of this soil. The Kirchberg-cambisol is a sandy soil, with > 50 weight % sand fraction.

NT-plots show higher bulk density and lower total porosity than conventionally tillaged (CT) plots and is due to a natural settlement of particles free from tillaging practices. As a consequence, there is a loss of coarse pores which leads to a slightly diminiushed available water capacity.The silty Loess-chernozem at Pixendorf is slightly alkaline with a medium carbonate content. The siliceous sites Pyhra and Kirchberg are weakly acicic. The contents of soil organic matter decrease with soil depth.

The Pixendorf-chernozem features a rather low content, the Pyhra-stagnosol a medium content and the Kirchberg-cambisol a high content of Fe-oxide, from this the expected sorption capacity for glyphosate and AMPA increased respectively from the chernozem, over the stagnosol to the higher weathered cambisol at Kirchberg.

At the first soil sampling immediately after field application of Round Up Max only a part of the applied glyphosate amount was detected in the upper 0-2 cm. The main part of the herbicide adheres at the green plant cover and at first does not enter the soil surface. Most of the applied glyphosate was transported and adsorbed in deeper horizons after the second and third time interval with concomitant increase of the AMPA content. The increase of AMPA 3 days after application of Round Up Max shows the very quick degradation of glyphosate to AMPA.

The results showed distinguished contents of glyphosate and AMPA in different soils at the same soil depth, according to their chemical-mineralogical adsorption properties.
Based on previous studies, we expected that glyphosate and AMPA would be more strongly adsorbed in soils with a higher iron-oxide (Fed) content (Mentler et al., 2007; Pessagno et al., 2008). Our results show that both the Kirchberg-cambisol and the Phyra-stagnosol, with a higher pedogenic iron-oxide content, adsorbed a distinctly higher quantity of glyphosate and AMPA than the Pixendorf –chernozem which had a distinctly lower iron-oxide content and a Kd-value about 10 times lower than that from Kirchberg (Klik et al., 2010).

Moreover, the weakly weathered Pixendorf-chernozem consequently shows a low content of amorphous (Feo) iron-oxides (973 mg Feo kg^{-1} soil) with respect to the more highly weathered Kirchberg- cambisol (3402 mg Feo kg^{-1} soil) and Pyhra-stagnosol (3279 mg Feo kg^{-1} soil). Higher contents of amorphous iron-oxides (Feo) (Piccolo et al., 1994; Gerritse et al., 1996), lead to higher sorption of glyphosate and AMPA, probably due to a larger and more reactive surface area of amorphous iron-oxides.

Thus, iron-oxides in general seem to be a key parameter for glyphosate and AMPA adsorption in soils. Our study confirmed this: the analysis showed lower contents of Fed and Feo for the Pixendorf-chernozem, with consequently lower adsorption of glyphosate and AMPA compared with the Kirchberg-cambisol and the Pyhra-stagnosol (see also the results of Eberbach, 1997).

Rainfall simulation experiments.

The results show that under normal practical conditions (e.g no rainfall is expected after application), the potential adsorption capacity of the Kirchberg-cambisol, with about 16.000 ppm pedogenic Fe-oxides, is confirmed compared to the lower adsorption capacity of the Pixendorf-chernosem (about 8.000 ppm pedogenic Fe-oxides).

Considering the enormous differences in the run-off amounts between the two sites Pixendorf and Kirchberg it can be concluded how important the soil surface conditions and vegetation cover of the agricultural fields for erosion risk and pollution risk of surface water are.

The Pixendorf-chernozem was less compacted and showed a higher permeability than the Kirchberg-cambisol. These parameters affect infiltration, fast flow and preferential flow processes in the soil and finally the glyphosate dispersion by run-off. In the rainfall simulation experiments under comparable simulation conditions, the amount of run-off at Kirchberg was approx. 10 times higher than at the Pixendorf site, due to the much better infiltration rate of the latter. Moreover, the total loss of glyphosate (NT+CT) through run-off was more than double on the Kirchberg site, which confirms the higher risk of pesticide pollution for surface waters from agricultural fields with high erosion intensity.

By the performed investigations only hypothetical conclusions can be stated about the fate of the „rest" glyphosate (e.g. not found in soils or run-off fractions). A part of glyphosate could remain adhering to the green plant organs through the adhesion substances added in Round Up Max. A part could have been leached to deeper soil horizons and than there adsorbed through the soil. A small loss out of the installed simulation plots through lateral fissures cannot be totally excluded.

Previous studies by Veiga *et al.*, (2001), Kjaer *et al.*, (2005) and Candela *et al.*, (2007), showed that adsorption processes for glyphosate can be strongly influenced by physical soil conditions like structure and water infiltration rate. This should be the aim of further investigations for a better understanding of the fate of glyphosate and AMPA in the environment.

On the other hand, Veiga *et al.*, (2001) demonstrated that even in some acidic soils with high levels of cations in the soil solution, e.g. Cu^{2+} and Al^{3+}, together with a high porosity and moderate permeability during the rainfall period, glyphosate could be transported vertically through the soil when it is quickly bound to the particulate suspended material.
This study and other investigations underline that the influence of soils on the behaviour and fate of glyphosate and AMPA is multifactorial and complex.

For a risk assessment of contamination of surface – and groundwater with glyphosate through erosion processes, the following overall conclusions temporary conclusions can be stated:
Glyphosate is not totally adsorbed in soils. The potential adsorpion capacity of soils depends on their physico-chemical-mineralogical-biological composition and is therefore site dependend.
This work could show that the definitive dissipation of glyphosate and AMPA through erosion is not only influenced by the conventional erosion parameter like soil erodibility, precipitation erosivity, slope, crop, as well as the soil composition, but finally also the soil structure state, e.g.compaction degree, crusting, infiltrability, pore size distribution and geometry, hydraulic conductivity.
The rain simulation experiments could clearly show that even a potentially high erodible site like the Pixendorf-chernozem suffers low damages, if adequate

protectioin practices (in this case the grow of a green crop after the wheat yield) are applied ; respectively that even a potentially high adsorbing soil like the Kirchberg-cambisol suffers strong erosion damages , if its structure is unfavourable. In this case in one of the NT -plot repetitions up to 47 % of the applied glyphosate amount were dispersed with run-off.

Traces of glyphosate in percolation water at Pixendorf, probably from previous conventional field application of Round Up Max, could confirm the general low adsorption capacity of chernozems from Loess and their risk of transport towards groundwater. Moreover, analysis of water from run-off retention basins in the landscape in the surrounding of Pixendorf confirmed that through high erosion processes, especially in corn crop, glyphosate is partly transported outside of the applicated agricultural fields.

5 References

Aubin, A.J. and Smith, A.E. 1992. Extraction of [^{14}C]Glyphosate from Saskatchewan soils. *J. Agric. Food Chem.* **40**, 1163-1165.

Barja, B.C. and dos Santos Afonso, M. 1998. An ATR_FTIR study of glyphosate and its Fe(III) complex in aqueous solution. *Environ. Sci.Technol.* **32**, 3331-3335.

Barja, B.C., Herszage, J., and dos santos Afonso, M. 2001. Iron(III)-phosphonate complexes. *Polyhedron* **20**, 1821-1830.

Baylis, A.D. 2000. Why glyphosate is a global herbicide: strengths, weaknesses and prospects. *Pest Manag Sci* **56**, 299-308.

Blum, W.E.H. 2007. Bodenkunde in Stichworten. Gebr. Borntraeger Verlagsbuchhandlung, Berlin, Stuttgart 2007.

Borjesson, E. and Torstensson, I. 2000. New methods for determination of glyphosate and (aminomethyl)phosphonic acid in water ands soil. *J. Chromatogr. A* **886**, 207.

Candela, L., Álvarez-Benedi, J., Condesso de Melo, M.T., and Rao, P.S. C. 2007. Laboratory studies on glyphosate transport in soils of the Maresme area near Barcelona, Spain: Transport model parameter estimation. *Geoderma* **140**, 8-16.

Carlisle, S.M., and Trevors, J.T. 1988. Glyphosate in the environment. *Water Air Soil Poll.* **39**,409-420.

Damonte, M.,Torres Sánchez, R.M., and dos Santos Afonso, M. 2007. Some aspects of the glyphosate adsorption on montmorillonite and its calcinated form. *APPL CLAY SCI*, **36**, 86-94.

da Cruz, L.H., de Santana, H., Bussamra, C.T., Zaia, V., and Zaia, D.A. M. 2007. Adsorption of glyphosate on clays and soils from Paraná State: Effect of pH and competitive adsorption of phosphate. *Braz. Arch. Biol. Technol.* **50 (3)**, 385-394.

Eberbach, P. 1997. Applying non-steady-state compartmental analysis to investigate the simultaneous degradation of soluble and sorbed glyphosate (N-(Phosphonomethyl)glycine) in four soils. *Pestic. Sci.* **52**, 229-240.

Gawlik, B.M., and Muntau, H. 1999. Eurosoils II Laboratory Reference Materials for Soil Related Studies. Joint Research Centre. European Commission. Official Publication of the European community.

Gerritse, R.G., Beltran, J., and Hernandez, F. 1996. Adsorption of atrazine, simazine and glyphosate in soils of the Gnangara mound, Western Australia. *Australian J. Soil Res.* **34**, 599-607.

Ghanem, A., Bados, P., Estaun, A.R., de Alencastro, L.F., Taibi, S., Einhorn, J., and Mougin, C. 2007. Concentrations and specific loads of glyphosate, diuron, atrazin, nonylphenol and metabolites thereof in French urban sewage sludge. *Chemosphere* **69**, 1368-1373.

Gimsing, A.L., Borggaard, O.K., Jacobsen, O.S., Aamand, J., and Sørensen, J. 2004. Chemical and microbiological soil characteristics controlling glyphosate mineralization in Danish surface soils. *Appl. Soil Ecol.* **27**, 233-242.

Haghani, K,, Salmanian, AH,, Ranjbar, B,, Zakikhan-Alang, K. and Khajeh, K. 2007. Comparative studies of wild type Escherichia coli 5-enopyruvylshikimate 3-phosphate synthase with three glyphosate-intensive mutated forms: Activity, stability and structural characterization. *Biochim. Biophys. Acta* 2007., doi: 10.1016/j.bbapap.2007.07.021

Hanke, I., Singer, H., and Hollender, J. 2008. Ultratrace-level determination of glyphosate, aminomethylphosphonic acid and glufosinate in natural waters by solid-phase extraction followed by liquid chromatography-tandem mass spectrometry: Performance tuning of derivatisation, enrichment and detection. *Anal. Bioanal. Chem.* **391 (6)**, 2265-2276. DOI:10.1007/s00216-008.2134-5.

Haslmayr, H.-P. 2010. "Rote Liste" schützenswerter Böden Österreichs: Eine Methode zur Definition von schützenswerten Bodenformen als Planungsgrundlage flächenwirksamer Landnutzungen. Dissertation, Universität für Bodenkultur Wien.

Hennion, M.C. 1999. Solid-phase extraction: Method development, sorbents, and coupling with liquid chromatography. *J. Chromatogr. A*, 3–54.

Hogendoorn, E.A., Hoogerbrugge, R., Baumann, R.A., Meiring, H.D., de Jong, A.P.J.M., and van Zoonen, P. 1996. Screening and analysis of polar pesticides in environmental monitoring programmes by coupled-column liquid chromatography and gas chromatography-mass spectometry. *J. Chromatogr. A*, **754**, 49-60.

Ibáñez, M., Pozo, O.J., Sancho, J.V., López, J., and Herándex, J. 2005. Residue determination of glyphosate and aminometylphosphonic acid in water and soil samples by liquid chromatography coupled to electrospray tandem mass spectrometry. *J. Chromatogr. A*, **1081**, 145-155.

Ibáñez, M., Pozo, O.J., Sancho, J.V., López, J., and Herándex, J. 2006. Re-evaluation of glyphosate determination in water by liquid chromatography coupled to electrospray tandem mass spectrometry. *J. Chromatogr. A*, **1134**, 51-55.

Kjaer, J., Olsen, P., Ullum, M., and Grant, R. 2005. Vadose zone processes and chemical transport: Leaching of glyphosate and amino-methylphosphonic acid from Danish agricultural field sites. *J. Environ. Qual.* **34**, 608-620.

Klik, A., Trümper, G., Baatar, U., Strohmeier, S., Liebhard, P., Deim, F., Moitzi, G., Schüller, M., Rampazzo, N., Mentler, A., Rampazzo Todorovic, G., Brauner, E., Blum, W.E.H., Köllensperger, G., Hann, S., Breuer, G., Stürmer, B., Frank, S., Blatt, J., Rosner, J., Zwatz-Walter, E., Bruckner, R., Gruber, J., Spiess, R., Sanitzer, H., Haile, T.M., Selim, S., Grillitsch, B., Altmann, D., Guseck, C., Bursch, W., und Führhacker, M. 2010. Einfluss unterschiedlicher Bodenbearbeitungssysteme auf Kohlenstoffdynamik, CO2-Emissionen und das Verhalten von Glyphosat und AMPA im Boden. Endbericht. Forschungsprojektnr.: 100069, GZ BMLFUW-LE.1.3.2/0130- II/1/2006, im Auftrag des BMLFUW in Kooperation mit den Bundesländern Niederösterreich und Steiermark. 299 S.

Krenn, A. 1999. Estimation of soil water retention using two models based on regression, Ananlysis and an aritificial neual network: Proceedings of the International Workshop on charakteriazation and measurement of the Hydraulik properties of unsaturated porous media, 22-24. October 1997, Riverside, California, 1261-1267.

Landry, D., Dousset, S., Fournier, J.-C., and Andreux, F. 2005. Leaching of glyphosate and AMPA under two soil management practices in Burgundy vineyards (Vosne-Romanée, 21-France). *Environ. Pollut.* **138**, 191-200.

Lee, E.A., Zimmerman, L.R., Bhullar, B.S., and Thurman, E.M. 2002. Linker-assisted immunoassay and liquid chromatography/mass spectrometry for the analysis of glyphosate. *Anal. Chem.* **74**, 4973.

Mamy, L., Barriuso, E. and Gabrielle, B. 2005. Environmental fate of herbicides trifularin, metazachlor, metamitron and sulcotrione compared with that of glyphosate, a substitute broad spectrum herbicide for different glyphosate-resistant crops. *Pest. Manag. Sci.* **61**, 905-916.

McConnell, J.S. and Hossner, L.R. 1985. pH-dependent adsorption isotherms of glyphosate. *J. Agric. Food Chem.* **33**, 1075-1078.

Mentler, A., Paredes, M., and Fuerhacker, M. 2007. Adsorption of glyphosate to cambisols, podzols and silica sand. *Alfa conference proceedings*, Salzburg.

Miles, C.J. and Moye, H.A. 1988. Extraction of glyphosate herbicide from soil and clay minerals and determination of residues in soils. *J. Agric. Food Chem.* **36**, 486-491.

Morillo, E., Undabeytia, T., and Maqueda, C. 1997. Adsorption of glyphosate on the clay mineral montmorillonite: Effect of Cu(II) in solution and adsorbed on the mineral. *Environ. Sci. Technol.* **31**, 3588-3592.

Morillo, E., Undabeytia, T., Maqueda, C., and Ramos, A. 2000. Glyphosate adsorption on soil of different characteristics: Influence of copper addition. *Chemosphere* **40**, 103-107.

Nestroy, O., Danneberg, O.H, Englisch, M., Gessl, A., Hager, H., Herzberger, E., Kilian, W., Nelhiebel, P., Pecina, E., Pehamberger, A., Schneider, W., Wagner, J., 2000: Systematische Gliederung der Böden Österreichs (Österreichische Bodensystematik 2000).- *Mitt. der ÖBG*, Heft 60.

Peruzzo, P.J., Porta, A.A., and Ronco, A.E. 2008. Levels of glyphosate in surface waters, sediments and soils associated with direct sowing soybean cultivation in north pampasic region of Argentina. *Environ. Pollut.* DOI: 10.1016/j.envpol.2008.01.015

Piccolo, A., Celano, G., Arienzo, M., and Mirabella, A. 1994. Adsorption and desorption of glyphosate in some European soils. *J. Environ. Sci. Heal.* **B 29**, 1105-1115.

Pessagno, R.C., Torres Sánchez, R.M, dos Santos Afonso, M. 2008. Glyphosate behavior at soil and mineral-water interferences. *Environ. Pollut.* **153(1)**, 53-59.

Rueppel, M.L., Brightwell, B.B., Schaefer, J. and Marvel, J.T. 1977. Metabolism and degradation of glyphosate in soil and water. *J. Agric. Food Chem.*, Vol. **25**, No. 3, 517-528.

Sancho, J.V., Hernández, F., López, F.J., Hogendoorn, E.A., and van Zoonen, P. 1996. Rapid determination of glufosinate, glyphosate and aminomethylphosphonic acid in environmental water samples using fluorogenic labeling and coupled-column liquid chromatography. *J. Chromatogr. A,* **737**, 75.

Sancho, J.V., Pozzo, O.J., López, F.J., and Hernández, F. 2002. Different quantitation approaches for xenobiotics in human urine samples by liquid chromatography/electrospray tandem mass spectrometry. *Rapid Commun. Mass Spectrom.* **16**, 639. DOI: 10.1002/rcm.617

Schnurer, Y., Persson, P., Nilsson, M., Nordgren, A., and Giesler, R. 2006. Effects of surface sorption on microbial degradation of glyphosate. *Environ. Sci. Technol.* **40**, 4145-4150.

Soulas, G., and Lagacherie, B. 2001. Modelling of microbial degradation of pesticides in soils. *Biol. Fertil. Soils* **33**, 551-557.

Strauss, P., Pitty, J., Pfeffer, M., and Mentler, A. 2000. Rainfall Simulation for outdoor Experiments. In : P. Jamet and J. Coirnejo (eds): Current research methods to assess the environmental fate of pesticides. pp 329-333, INRA Editions.

Strauss, P. (2008). Personal cvommunication.

Ternan, N.G-, Mc Grath, J.W., Mc Mullan, G. and Quinn, J.P. 1998. Review: Organophosphonates: occurance, synthesis and biodegradation by microorganisms. World of Journal of Microbiology & Biotechnology **14**, 635-647.

Thompson, D.G. 1989. Liquid chromatographic method for quantitation of glyphosate and metabolite residues in organic and mineral soil, stream sediments, and hardwood foliage. *J. Assoc. Off. Anal. Chem.* **72**, 2.

Veiga, F., Zapata, J.M., Fernandez Marcos, M.L., and Alvarez, E. 2001. Dynamics of glyphosate and aminomethylphosphonic acid in a forest soil of Galicia, north-west Spain. *SCI TOTAL ENVIRON* **271**, 135-144.

Vreeken, R.J., Speksnijder, K., Bobeldijk-Pastorova, I., and Noij, H.M. Th. 1998. Selective analysis of the herbicide glyphosate and aminomethyl phosphoric acid in water by on-line solid-phase extraction-high-performance liquid chromatography-electrospray ionization mass spectrometry. *J. Chromatogr. A*, **794**, 187-199.

WRB 2006. World reference base for soil resources 2006. *World Soil Resources Reports*, **No.103**. FAO, Rome.

Zaranyika, M.F., and Nyandro, M.G. 1993. Degradation of glyphosate in the aquatic environment: An enzymatic kinetic model that takes into account microbial degradation of both free and colloidal (or sediment) particle adsorbed glyphosate. *J. Agric. Food Chem.* **41**, 838-842.

Zhou, D.-M., Wang, Y.-J., Cang, L., Hao, X.-Z., and Luo, X.-S. 2004. Adsorption and co-sorption of cadmium and glyphosate on two soils with different characteristics. *Chem.* **57**, 1237-1244.

Acknowledgments:

Bundesministerium für Land- und Forstwirtschaft, Umwelt und Wasserwirtschaft (BMLFUW), Amt der Niederösterreichischen Landesregierung, Amt der Steiermärkischen Landesregierung, for the financiation of the research project "EDISSOC" Nr.: 100069, GZ BMLFUW-LE.1.3.2/0130-II/1/2006;

Landwirtschaftliche Fachschulen Tulln, Mistelbach, Pixendorf, Pyhra and Kirchberg am Walde for providing the field experimental plots;

o.Univ.Prof.Dipl.-Ing.Dr.DDr.h.c.mult.Dr. Winfried E.H. Blum for his supervision and continuous support during the whole working time;

Ao.Univ.Prof.Dipl.-Ing.Dr. Eduard Klaghofer for his accurate review oft he manuscript;

Ao.Univ.Prof.Dipl.-Ing.Dr. Peter Liebhard for his accurate review oft he manuscript;

Ass.Prof.Dipl.-Ing.Dr. Axel Mentler for his endless help, support and encouragement in the laboratory during the development of the new extraction method;

Dipl.-Ing. Dr. Peter Strauß und Dipl.-Ing Alexander Eder, IKT Petzenkirchen, for their great help in providing and co-executing the field rainfall simulation experiments;

AmtDir. Ing. Ewald Brauner and the whole laboratory staff of the Institute of Soil Research, Department of Forest- and Soil Sciences, University of Natural Resources and Life Sciences, Vienna for the great support in the lab;

Curriculum vitae
of
Gorana Rampazzo Todorovic

	Name: **Gorana Rampazzo Todorović** **Place and date of birth:** Pančevo, Serbia, 18-12-1971 **Marital status:** married, 1 child **Nationality:** Serbian **Adresse:** Neusiedlerstrasse 62/1, Mödling, Austria **E – mail:** gorana.todorovic@boku.ac.at goranatodorovic@hotmail.com
Education	
1978 – 1986	Elementary school, Pancevo, Serbia.
1986 - 1990	Secondary school for Natural Sciences, school-leaving-examination, Serbia.
1990 - 1997	Diploma Study of Soil Sciences, Faculty of Agriculture, University of Belgrade, Serbia. Diploma thesis: *"Waters protection in agricultural used soils"*.
1997 - 1999	MSc in Soil Science, Department for Soil Sciences, Faculty of Agriculture, University of Belgrade, Serbia (interrupted).
2002 (3 months)	Advanced training course "Methods for Environmental Analysis in Landspace" at Centro Sperimentale per lo Studio e l'Analisi del Suolo (CSSAS), Department for Agroecological Sciences and Technologies, Faculty of Agriculture, University of Bologna, Italy.
2004-2005	MSc in *'Natural Resources Management'*, Institute for Soil Research, Department of Forest- and Soil Sciences, University of Natural Resources and Life Sciences, Vienna. Master thesis :"Influence of rhizosphere microbes on heavy metal uptake by *Salix caprea* and *Thlaspi goesingense* "
2007-2012	PhD in Soil Sciences at the Institute for Soil Research, Department of Forest- and Soil Sciences, University of Natural Resources and Life Sciences, Vienna. PhD thesis: " Behavior of glyphosate and AMPA in soils under the influence of different tillage systems and erosion"
Professional experiences	Assistant at the Department for Soil Sciences, Faculty of Agriculture, University of Belgrade, Serbia.
1997 - 1999	Co-work at following projects of Ministry of Science and Technology, Serbia: 1. Soil parameters in Serbian soils. 2. Chemical and mineralogical properties of soils of the regions Sumadija and Pomoravlje, Serbia. 3. Experiments with new cultivation methods of soja and mais as second culture.
1999-2000	Co-work in a perfumery, Belgrade, Serbia.

2001	Co-work at «Glečer» - Society for Civil Engineering, Horticulture and Management of Public Green Areas.
2002 -2004	Co-work at research project at the University of Belgrade „Soil degradation and -contamination of agricultural soils in the surrounding area of Pancevo, Vojvodina". Cooperation with CSSAS of the University Bologna on: - GIS-application for pedological-geological aims. - Soil mapping and -classification.
2004 -2005	Co-work of literature study, field experiments, data collection at the Institute for Soil Research, Department of Forest- and Soil Sciences, University of Natural Resources and Life Sciences, FWF-projects: - "Microbial interactions in the rhizosphere of metal hyperaccumulating and hypertolerant plants in ultramafic soils: Characterization, functional relations and bioavailablity". - "Genomics for a better environment: molecular mechanisms involved in the metal accumulation by *Salix caprea* and associated microorganisms".
2005-2006	Co-work at EU-project no. 505428 (GOCE) „AQUA-TERRA: „Integrated Modelling of the river-sediment-soil groundwater system; advanced tools for the management of catchment areas and river basins in the context of global change" at Institute fof Soil Research, Department of Forest- and Soil Sciences, University of Natural Resources and Life Sciences, Vienna. Contract with Federal Agency for Water Management, Petzenkirchen, Lower Austria on: „Verification of automatically generated model parameters for detection of soil loss through water in Austria"
2006	Contract with Office of Upper Austrian Government, Department Environment and Equipments Techniques on „Compost Application Long term Experiment at the LBFS Ritzlhof 1993-2003" Compilation of the final report. Contract with the Institute for Physics and Material Sciences, Department of Material Sciences and Process Technology, University of Natural Resources and Life Sciences, Vienna, on: „Test of wood samples".
2006	Contract with Dr. M. Stemmer, AGES, on: „Application and extension of computer-supported Carbon-models (e.g. RothC) for estimation and validation of Carbon-Pools in 2 Austrian long term experiments". Contract with the Federal Agency of Research and Education for Forest, Natural Risk and Landscape, on: „Soil physical analyses of soil samples within the EU-project BioSoil".
2006-2008	Contract with the Institute for Soil Research, Department of Forest- and Soil Sciences, University of Natural Resources and Life Sciences, Vienna, on: "Compilation of a project proposal with cooperation of international partners for the submission to the ERA NET-call „SNOWMAN"„. Employed by the company VITANA- Salat- und Frischeservice GmbH. as a soil expert for nitrate problematic and quality control of nitrate and pesticide content in different vegetables.
2007-2010	Employed as researcher at the Institute for Soil Research, Department of Forest- and Soil Sciences, University of Natural Resources and Life Sciences, Vienna, on the project EDISSOC, Project Nr. 100069/3 „Effects of different soil management systems on carbon sequestration, CO_2 emissions and behaviour of glyphosate and AMPA in soils. Contract with the Institute for Soil Research, Department of Forest- and Soil Sciences, University of Natural Resources and Life Sciences, Vienna, on: "Viniculture in times of climate change". Partecipation in COST 639 action on on evaluation of performance of different

	soil-carbon models for Kyoto-reporting. Works on modelling of carbon stocks in soil in SoilTrEC EU-Project.
2010-2012	Maternity leave
2012	Employed as researcher at the Institute for Soil Research, Department of Forest- and Soil Sciences, University of Natural Resources and Life Sciences, Vienna, on the project SOILTREC, Project Nr. 7911008000 „Soil Transformations in European Catchments.
Languages	Serbo-Croatian (mother language) English (fluently) Italian (fluently) German (fluently)
Informatics knowledges	Microsoft Office Software CAD : Autocad (Autodesk) GIS : ArcView GIS (ESRI Inc.), IDRISI (Clarc University)
Private interests	painting, chorus singing, piano playing history, history of arts, anthropology cooking, gardening, animal care in zoos diving, snorkelling, sailing, trekking

i want morebooks!

Buy your books fast and straightforward online - at one of world's fastest growing online book stores! Environmentally sound due to Print-on-Demand technologies.

Buy your books online at
www.get-morebooks.com

Kaufen Sie Ihre Bücher schnell und unkompliziert online – auf einer der am schnellsten wachsenden Buchhandelsplattformen weltweit! Dank Print-On-Demand umwelt- und ressourcenschonend produziert.

Bücher schneller online kaufen
www.morebooks.de

VDM Verlagsservicegesellschaft mbH
Heinrich-Böcking-Str. 6-8   Telefon: +49 681 3720 174   info@vdm-vsg.de
D - 66121 Saarbrücken   Telefax: +49 681 3720 1749   www.vdm-vsg.de

Printed by Books on Demand GmbH, Norderstedt / Germany